Weather Patterns and Phenomena

A Pilot's Guide

Other Books in the TAB Practical Flying Series

The Pilot's Guide to Weather Reports, Forecasts & Flight Planning *by Terry T. Lankford*

Aviator's Guide to Navigation—2nd Edition *by Donald J. Clausing*

Stalls & Spins *by Paul A. Craig*

The Pilot's Air Traffic Control Handbook—2nd Edition *by Paul E. Illman*

The Pilot's Radio Communications Handbook—4th Edition *by Paul E. Illman*

Mastering Instrument Flying—2nd Edition *by Henry Sollman with Sherwood Harris*

Flying in Adverse Conditions *by R. Randall Padfield*

Avoiding Mid-Air Collisions *by Shari Stamford Krause, Ph.D.*

Weather Patterns and Phenomena

A Pilot's Guide

Thomas P. Turner

TAB Books
Division of McGraw-Hill, Inc.
New York San Francisco Washington, D.C. Auckland Bogotá
Caracas Lisbon London Madrid Mexico City Milan
Montreal New Delhi San Juan Singapore
Sydney Tokyo Toronto

pbk 1 2 3 4 5 6 7 8 9 FGR/FGR 9 9 8 7 6 5 4
hc 1 2 3 4 5 6 7 8 9 FGR/FGR 9 9 8 7 6 5 4

Library of Congress Cataloging-in-Publication Data

Turner, Thomas P.
Weather patterns and phenomena : a pilot's guide / by Thomas P. Turner.
p. cm.
ISBN 0-07-065601-0 ISBN 0-07-065602-9 (pbk).
1. Meteorology in aeronautics. I. Title.
TL556.T87 1994
629.132'4—dc20 94-14289
CIP

Acquisitions editor: Jeff Worsinger
Editorial team: Bob Ostrander, Executive Editor
Sally Anne Glover, Editor
Elizabeth J. Akers, Indexer
Production team: Katherine G. Brown, Director
Susan E. Handsford, Coding
Jan Fisher, Desktop Operator
Designer: Jaclyn J. Boone

PFS
0656029

Contents

PART 3 REGIONAL WEATHER

PART 4 FLIGHT PLANNING

PART 5 [illegible]

Acknowledgments

I DEPEND HEAVILY ON A HOST OF METEOROLOGISTS, WEATHER BRIEFERS, pilots, and authors for my ever-growing grasp of aviation weather. Every time I read an article, watch a broadcast, speak with pilots, or talk to Flight Service, I learn something new that will make me safer and improve my ability to teach weather to pilots. In these acknowledgments, I know I'll leave somebody out; I hope they'll understand, but I want to thank the major influences I've had for the general body of weather knowledge they've passed on to me. In no particular order, thanks to:

Richard L. Collins of *Flying*, then *AOPA Pilot*, and then *Flying* magazine again. Read anything he writes about weather or flying in general!

Norman Schuyler of *Private Pilot* and Thomas A. Horne of *AOPA Pilot*. Their columns served as my primary references for testing and expanding my weather knowledge.

Robert N. Buck, author of *Weather Flying*, the classic text on meteorological risk management.

Terry T. Lankford, author of *Pilot's Guide to Weather Reports, Forecasts and Flight Planning*.

Advisory Circular 0045B, Aviation Weather Services, for assistance in decoding the hieroglyphics of weather reporting.

R.D. Campbell, Donald J. Clausing, and Paul A. Craig, whose books filled in bits and pieces of the way I evaluate weather.

The fabulous folks who bring us The Weather Channel and A.M. Weather, not only for the information they convey but also for the opportunity for me to constantly test my level of weather development knowledge.

Flight Service and Flight Watch staffers everywhere, as well as those who provide DUAT and facsimile-transmitted weather information.

Each and every client whom I've had the pleasure of working with at the Beech Learning Center of Flight Safety International; you'll never know how much the "teacher" has learned from you.

Jeff Worsinger, acquisitions editor; Sally Glover, my patient editor; the editorial review board and anyone else at TAB/McGraw Hill who had confidence in me and was willing to invest the time it took to develop a legible product.

My wife Peggy and son Alan, for allowing the time it took to write this tome, as well as for suffering the incessant tap-tap-tapping of the word processor at all hours of the day and night.

Introduction

HISTORICALLY, WEATHER FACTORS CONTRIBUTE TO 25 PERCENT OF ALL general-aviation accidents. Nearly 40 percent of all fatal general-aviation crashes are the result of weather hazards, according to the Aircraft Owners and Pilots Association's Air Safety Foundation. Whether you're flying IFR across the country or VFR to the local fly-in hamburger stand, thoroughly understanding how and when weather hazards develop and move is critical to flight safety.

Yet very little emphasis is placed on weather knowledge in most pilot education programs. The cliche is true: you don't even have to answer a single weather-related question on the written airmen's exams in order to pass. You can earn all ratings through the ATP without ever correctly answering a weather question. Consequently, many ground-school programs skimp on meteorology in deference to aerodynamics, navigation, and a host of other important tasks. Perhaps unwittingly, the "system" contributes to the lack of weather awareness that leads to the large number of weather-related mishaps.

I'm not a meteorologist. Reading this book can't make you one either. What it can do is point out some of the general rules of weather development that can help you look at required weather briefings critically and plan your flight to limit the weather-related risk as much as possible. Meteorology is an inexact science, and many variables aren't fully understood. You need to have some basic knowledge to evaluate the weather in terms of the changing risk it presents.

What do you as a pilot need to know about weather in order to fly safely? You need to know something about general weather theory: what causes weather hazards to form, and how they move and develop after formation. You need to be able to anticipate the hazards likely to be found in a cold front, for instance, when your weather briefing reports one along your route. You have to be able to predict where and what weather signs you'll see en route, to verify or refute the forecast, and to determine what different hazards you might encounter if conditions you actually find differ from those expected. Part 1 of this book, Weather Theory, is designed to give you the knowledge to interpret the information you receive from Flight Service, DUAT, or some other approved outlet in terms of what weather phenomena you'll likely see. You can then compare this information with actual en route weather to determine the forecast's validity.

Of course, there are certain hazards beyond the capability of airplanes and pilots. Part 2, Aviation Weather Hazards, outlines the four classes of aviation weather hazards: thunderstorms, turbulence, reduced visibility, and ice. For each class of hazards, we'll cover the elements required for their formation, their life cycles, and the specific threat to airplanes that each hazard represents. We'll discover which aviation weather products (charts, etc.) point to the existence and severity of each hazard. We'll also look at many products normally provided in standard briefings, and a few given only for the asking. These can enable you to more precisely predict a hazard. We'll look at techniques for hazard avoidance, and what to do if you find yourself engulfed in the hazard's fury. The intent of Part 2 is to help you make a decision about whether to make a flight at all, and to pick a route and altitude designed to minimize your exposure to weather risk.

Part of the enjoyment of flying is going to new and unfamiliar places. Geographic regions experience unique weather patterns, however, that affect you despite your lack of familiarity. Part 3, Regional Weather, gives you a little insight into those geographic weather peculiarities that locals know. Knowledge of these will make you safer when flying to new locales. We'll divide the continental United States into 11 regions and examine some local weather patterns for each.

Part 4, Flight Planning, shows how to use all this knowledge to make an informed choice of routes and altitudes to minimize risk. From the time you first decide to make a flight until the point where your wheels make contact on landing, you need to constantly evaluate weather in terms of the hazards present or forecast, and you need to actively work to verify or refute your weather briefings to avoid a weather-related accident. We'll take some sample trips in challenging conditions to show how you can make safety-related weather decisions before takeoff and along the way.

Weather-reporting technology is changing rapidly. The Postscript looks briefly at some up-and-coming weather-observing devices that promise to make forecasts more and more accurate and herald the era of "nowcasting," minute-by-minute weather updates that precisely report changing conditions.

Understanding weather lets you realize the full utility of general aviation as a family recreation or business tool. If you know how to make informed weather-related decisions, not only will you avoid the cause of a quarter of all accidents and nearly half of all fatal mishaps, but you'll also make flying more reliable and enjoyable for you and your passengers.

PART 1

Weather theory

1
Air masses

WHAT CAUSES WEATHER? IF YOU HAD TO TRACE ALL WEATHER PHENOMENA back to a single source, what would it be? Temperature? Moisture content? Air pressure? Actually, you can reduce even those three factors to a single source, radiation from the sun.

TEMPERATURE

The earth's atmosphere reflects back, on average, about 55 percent of the radiation coming in from the sun. The remaining energy enters the atmosphere but does not warm the air from above. Instead, radiation passes through to the surface of the earth, which absorbs the sun's energy as warmth; that warmth transfers by contact with the lower atmosphere from the bottom up. It's like a microwave oven. Turn on the radiation, and the container doesn't get hot; the food inside does. If you leave the food in the container long enough, the container will get hot from contact with the food. This process, where surface heating is transferred to the air above, is called *terrestrial radiation*.

Different types of surfaces reflect or absorb the sun's radiation at varying rates. Water, for instance, reflects back most of the radiation, so air over wet, snowy, or vegetated areas tends to be relatively cool. Dark surfaces, such as parking lots and plowed, unseeded earth, on the other hand, absorb most of the incoming radiation and get hot; the air over these areas is relatively warm. Through terrestrial radiation, blocks of air take on the temperature characteristics of the underlying surface.

The amount of solar radiation that enters the atmosphere over a given point varies on a daily cycle. Radiation increases steadily from sunup through noon, decreases through the afternoon, and shuts off completely at night. The rate at which heat is lost from the surface through terrestrial radiation, however, is essentially fixed. What this means is that the air over a surface will heat and cool on a cycle called *diurnal variation*. Unless a weather system blows in to change this cycle, the hottest part of the day comes a few hours after local noon, when the sun has had the greatest amount of time to heat the ground but has not yet retreated significantly toward the horizon.

After sunset, the air will cool continuously during the night until it reaches its coolest point at or shortly after dawn, before the next day's warm-up begins (Fig. 1-1). Clouds, vegetation, or snow on the surface might limit the amount of temperature swing or the smoothness of the temperature transition; water takes more energy to change temperature one degree than just about any known substance, so moist climates, where there's a lot of water vapor in the air, do not heat as much or cool as much as drier climes. In Florida, for instance, there's often less than 10 degrees difference between the hottest and coolest parts of the day, whereas a dry climate such as New Mexico might have a diurnal variation of 50 degrees Fahrenheit or more (Fig. 1-2).

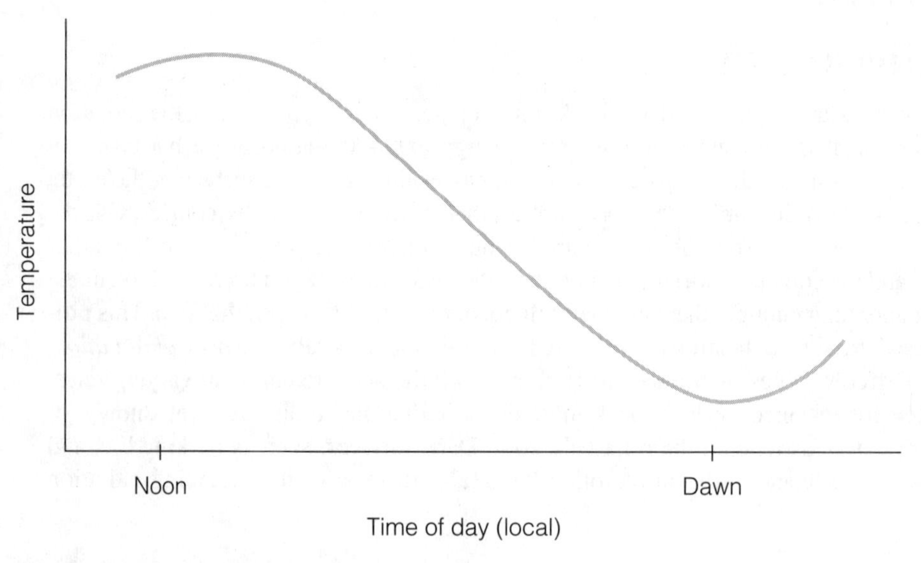

Fig. 1-1. *Typical diurnal temperature variation.*

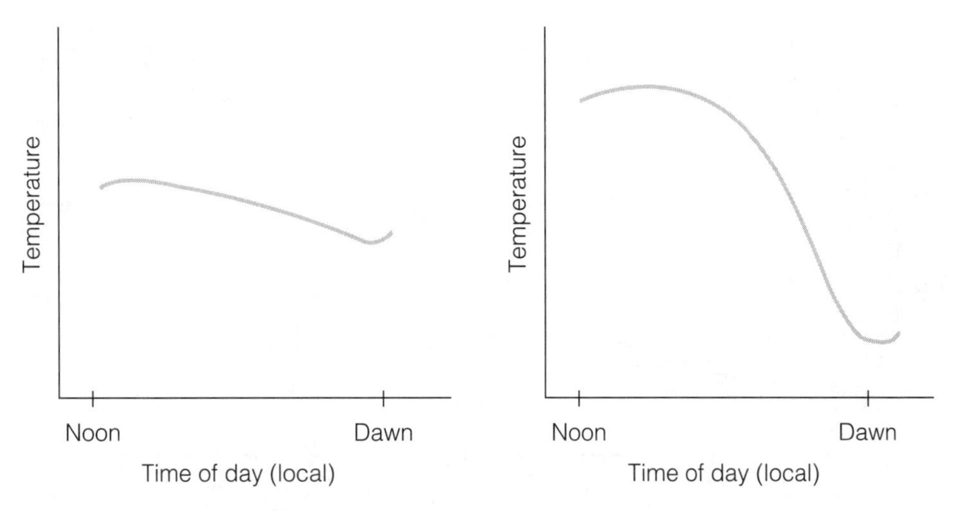

Fig. 1-2. *Typical diurnal variation in moist climates (left), and more arid regions (right).*

MOISTURE

As air heats up, moisture evaporates from bodies of water. This moisture can exist in gaseous or vapor form, or it might be frozen, suspended crystals of ice. The warmer the air, the more moisture it can contain. In fact, air at 70 degrees Fahrenheit can hold up to four times as much moisture as air at the freezing point. As it does with temperature, the air takes on the moisture characteristics of the surface beneath it. Air over the ocean tends to be moist, and air over land tends to be dry.

There are two mutually exclusive ways to state the amount of moisture contained in the air. One is *relative humidity*. Relative humidity is a percentage of the total amount of moisture the air can contain at its current temperature. Air with a relative humidity of 70 percent, for instance, is 7/10ths "full" of water. If the relative humidity is 100 percent, the air is saturated; add any more moisture and it will begin to condense into clouds, fog, or precipitation.

Because the ability of the air to contain moisture varies with the temperature, the relative humidity will change with the cycle of diurnal temperature swings. Unless a new weather system blows drier or moister air into an area, relative humidity will be at its lowest during the hottest part of the day, and relative humidity will rise as the air temperature drops. If there's enough moisture in the atmosphere that the relative humidity reaches 100 percent, condensation in the form of fog, low clouds, dew, or frost is likely. You can see, then, why fog is most likely to develop at or shortly after dawn, when temperatures are at their coolest, and why fog typically "burns off" a few hours later (Fig. 1-3).

Another way to express the amount of moisture in the air, the method commonly used in aviation weather reports, is the *dew point*. The dew point is the temperature to which the air must cool for the moisture already in it to reach saturation. In other

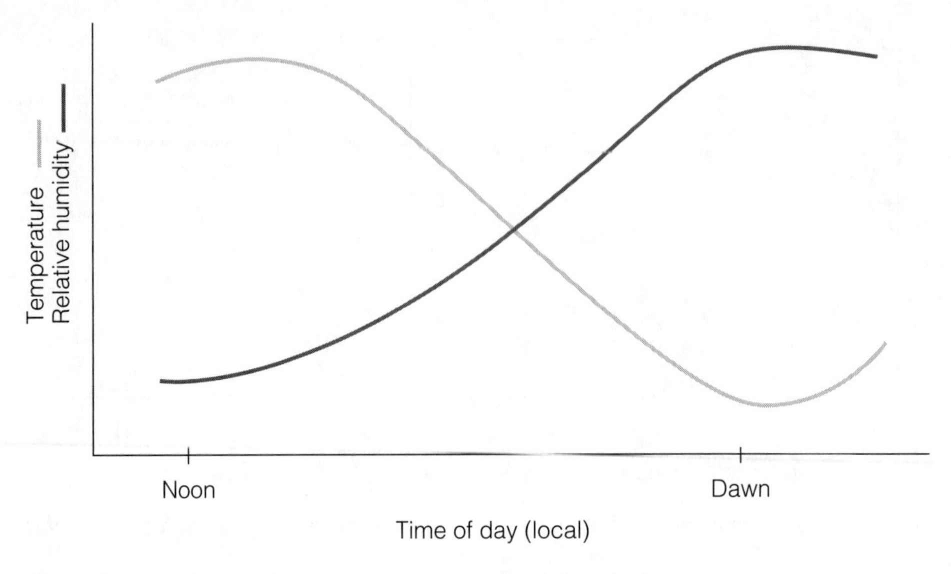

Fig. 1-3. *Relative humidity as it compares to diurnal variation.*

words, if the dew point is 56 degrees, the relative humidity would be 100 percent if the air cooled to 56 degrees. Because the amount of moisture in a parcel of air is essentially fixed, the dew point typically remains fairly steady over a day's time unless more moisture is added from somewhere. If the air temperature cools to near or below this dew point, condensation will occur. At altitude this forms a cloud; at the surface we call it fog. It's easy to see, again, that fog tends to develop at or shortly after dawn, when the air is coolest (Fig. 1-4). The *temperature/dew point spread* is the difference between the air temperature and the dew point. If the spread is within about 5 degrees Fahrenheit and narrowing, anticipate reduced visibility.

Let's put some of this knowledge to use. Let's say you're planning a flight and call to get a weather briefing. The Flight Service briefer tells you that there is a 4-degree spread between the temperature and the dew point, and issues the warning that "VFR flight is not recommended" or "low IFR conditions are expected" because of the possibility of reduced visibility. Visibilities are currently good. Should you try the flight anyway? It depends. You can predict a lot knowing about the diurnal temperature cycle and the time of day. If you had called between late afternoon to just before or around sunrise, for instance, you know that the temperature is likely to drop another few degrees before the day's warming begins. In this case, the temperature/dew point spread will probably narrow, and limited visibility is likely. If the sun has already been up for an hour when you receive the report, however, the temperature should be climbing, and visibility should remain good despite the Flight Service warning. You can ask the briefer to give you the temperature

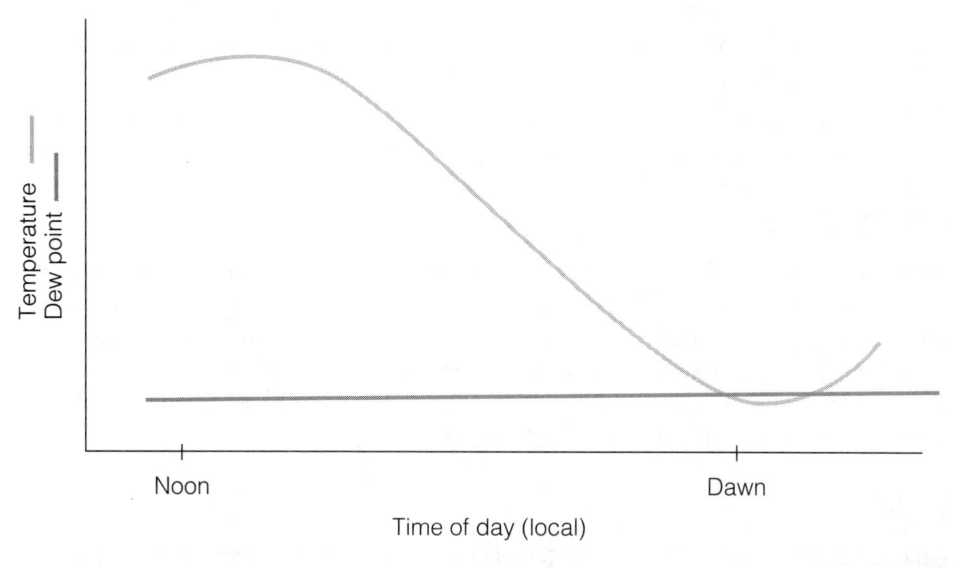

Fig. 1-4. *Dew point as it compares to diurnal variation.*

trend over the last several hours to see whether the temperature/dew point spread is widening or narrowing.

Be cautious about overgeneralizing, however. Ask yourself another question before you make the go/no-go decision: is something happening to add moisture to the air? Is the wind blowing strongly from a source of moisture, such as a large lake or river, or the ocean? That might add water vapor to the air, narrowing the temperature/dew point spread and causing a reduction in visibility. Is there a lot of water on the ground left over from a heavy rain the night before? If so, it will begin to evaporate into the air when the sun strikes it. Air over rain-soaked fields, or layers of snow when the temperature is near or slightly above freezing, increases in moisture content when the warming begins, increasing the humidity.

Here, then, are some rules of thumb about temperature, moisture content, and the likelihood of reduced visibility development:

- Temperature swings in a given weather system are roughly cyclic, making a prediction of improving or worsening temperature/dew point spreads possible.
- If a check of the weather in the late evening, such as by watching the late local news on television, mentions a relative humidity higher than around 80 percent and/or a temperature/dew point spread within about 5 degrees, anticipate fog or poor visibility around dawn the next day. Plan to depart before sunup to avoid the restriction in visibility (after deciding whether you want to overfly possible areas of fog), or anticipate waiting several hours for the skies to clear.

- When poor visibility is forecast, you can make a better-informed go/no-go decision by knowing the time of day and asking the weather briefer for the temperature and dew point trends over the last few hours.
- Before making the final decision to fly, ask yourself if there's anything adding moisture to the air, such as evaporating water or snow, or wind off a body of water.

AIR MASSES

So far we've talked about characteristics of temperature and moisture content, and how blocks of air take on these characteristics. These blocks of air are called *air masses*, which, on a global, continental, or even local scale, have roughly consistent properties of temperature and moisture content. Most observed weather phenomena occur at the conflict of dissimilar air masses. How, then, do these air masses come to meet? It's through the force of the wind.

WIND

Wind is another end result of atmospheric heating. Let's say we have a mass of air over two surfaces: over a body of water, and over an inland area (Fig. 1-5). The sun begins

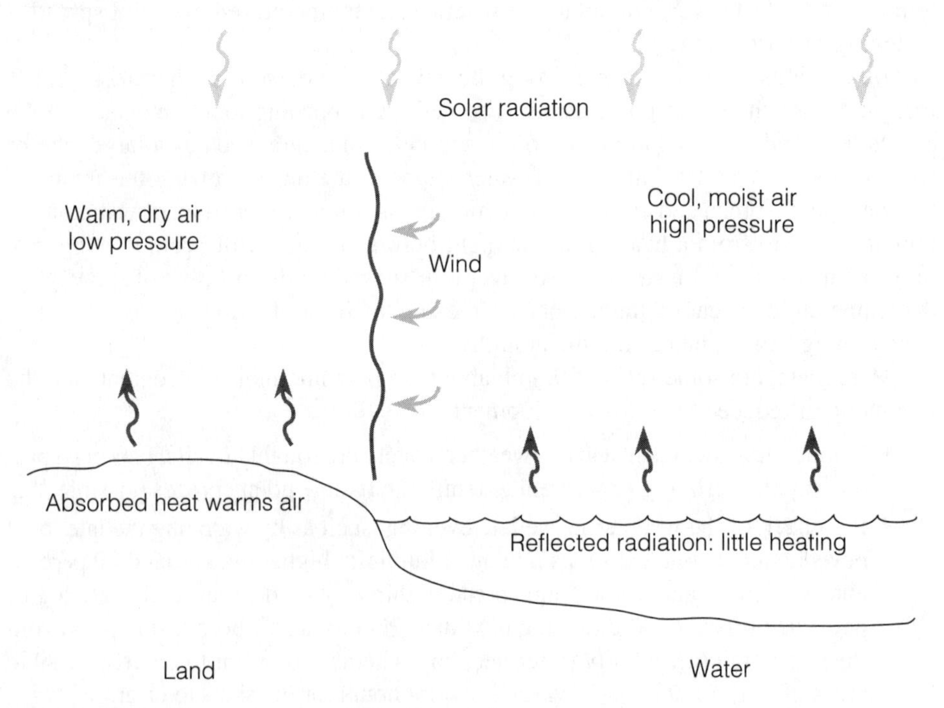

Fig. 1-5. *Wind formation over dissimilar surfaces.*

to beat down on the surfaces, but the water reflects back a lot of the energy. The drier land more rapidly heats the air above it, while the wetter air over water remains relatively cool. Now we have two separate air masses, one relatively warm and dry, the other relatively cool and moist. What happens to a gas, such as the air, when it heats up? It expands, and its pressure drops. Over the warm, inland area, then, we have air pressure lower than that over the water, and the warmer air rises, creating a void in the lower atmosphere. The cooler, denser air, which tends to descend, flows in to fill this partial vacuum; wind now flows from the moister air mass into the drier one. The bigger the temperature difference, and therefore the greater the contrast in atmospheric pressure, the stronger the wind flow. You can see that air is always going to flow away from areas of high pressure and into areas of low pressure. We'll look at some other general properties of wind movement in chapter 2.

This moist, cool air continues to collect in the area of low pressure, increasing the humidity in the low until perhaps the temperature and dew points meet. This is when condensation occurs, and visible moisture precipitates out in the form of clouds, rain, snow, or fog. High pressure areas, then, tend to repel adverse weather, and moisture funnels into low pressure areas, creating most weather hazards.

CONDENSATION

As stated earlier, condensation occurs when the temperature and dew points meet, or when the relative humidity equals 100 percent. Adding moisture to an air mass through the force of wind, as we've seen, can cause condensation; cooling of an already moist air mass is another likely cause of condensation. This cooling can be brought on by contact with colder air, such as along the boundaries of an air mass, by the flow of moist air over a cooler surface, or by expansion, when air is forced upward in the atmosphere by convection, frontal activity, or an upslope wind flow. Whether the moisture becomes a cloud, mist, or fog depends on the altitude and moisture content; if the particles of water become too heavy for the atmosphere to suspend them, they fall out as rain, ice, or snow, depending on the temperature of the air.

It would be great to be able to say that, for instance, a temperature/dew point spread of 2 degrees means a visibility of 2 miles, and such, but that's not the case. Condensation doesn't occur in a vacuum; instead, some *condensation nuclei*, such as particles of dust or smoke in the atmosphere, must be present for condensation to occur. This is why a 3-degree temperature/dew-point spread in Kansas might mean 10-mile visibility, while the same conditions in New Jersey, typically plagued by much greater levels of atmospheric pollution, might reduce visibilities to less than a mile. Wind also plays a factor; higher wind speeds tend to limit condensation and improve visibility.

STABILITY

What happens to moisture that condenses is largely a matter of atmospheric stability. As we'll see, temperature generally drops consistently with an increase in altitude. If warm air manages to override colder air, or if the rate of temperature loss is less than normal, any updrafts created by surface heating or other factors don't cool as fast as the still air around them; they continue to accelerate upward. We call this an *unstable air mass*. Instability is one cause of turbulence and is a required element for the formation of storms.

Stable air masses, on the other hand, tend to cancel out any lifting action, generally leading to more favorable flying conditions. Nothing's perfect, however; stability tends to reduce visibilities if moisture is present, and this is one part of the formation of a mountain wave.

As a rule, warm air masses and areas of relatively low air pressure are unstable, while cooler, higher pressure areas tend to be stable. We'll look at stability more closely in the chapters on clouds and fronts.

Here's a review of some tenets of air mass theory:

- Wind is the result of uneven heating of the air and the subsequent differences in air pressure. The greater the temperature and pressure contrast, the stronger the wind.
- Wind flows from areas of high pressure into areas of lower pressure.
- Most adverse weather is associated with low pressure areas; high pressure areas tend to repel bad weather.
- Condensation occurs when moisture is added to an air mass until saturation is reached, or if a moist air mass cools due to contact with colder air or surfaces, or expansion due to lifting in the atmosphere.
- Condensation occurs when: moisture is added to an air mass until saturation is reached, a moist air mass cools due to contact with colder air or surfaces, or a moist air mass cools as it rises in the atmosphere.
- The form and extent of condensation depends on the altitude, the temperature, the moisture content, stability, and the presence of condensation nuclei.
- Warm air masses of low pressure tend to be unstable, while high-pressure systems, created by relatively cooler air, are typically stable.

2
The atmosphere

KNOWING A LITTLE ABOUT THE STRUCTURE OF THE ATMOSPHERE WILL HELP you understand more about weather patterns.

LEVELS OF THE ATMOSPHERE

There are two levels of the atmosphere that concern you as a pilot (Fig. 2-1). The lowest level of the atmosphere, the *troposphere*, contains about 95 percent of all the earth's air. It's close enough to the surface of the earth to be affected by terrestrial radiation; the primary characteristic of the troposphere is that there is a net loss of air temperature with an increase in altitude. Virtually all weather phenomena occur in this lowest layer.

Eventually, altitude becomes great enough that there is little change in temperature; the air is too high to be warmed by terrestrial radiation. This layer of the atmosphere is called the *stratosphere*, where there is a roughly consistent temperature regardless of altitude. The stratosphere consists largely of clear, smooth air.

What we have, then, are two dissimilar air masses, the relatively warm, moist troposphere, and the cool, dry stratosphere. As with any other contrasting air masses, there is a boundary of adverse weather at the point of contact. This boundary between the troposphere and the stratosphere is called the *tropopause*.

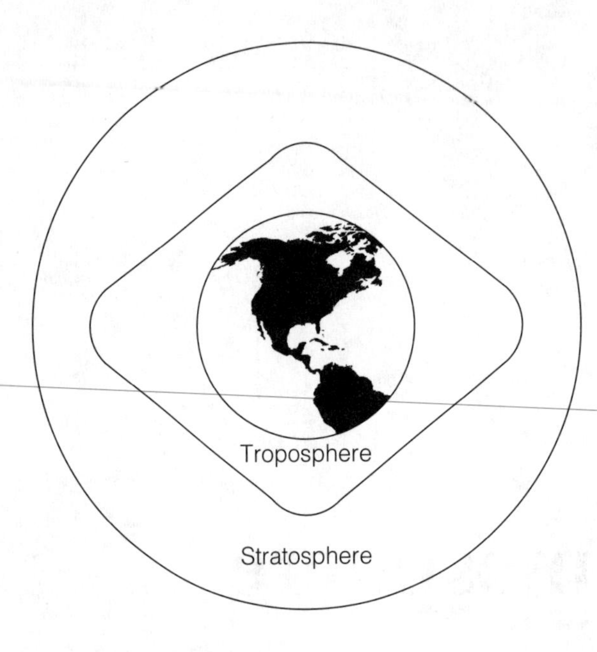

Fig. 2-1. *Levels of the atmosphere that affect pilots.*

The tropopause is a layer of generally high-speed wind and turbulence. This is where the jet stream, a high-speed core or river of air, occurs. We'll talk more about the jet stream in chapter 5. The tropopause height varies seasonally and geographically, averaging 60,000 feet over the equator, sloping to 25,000 feet over the poles, and dropping to half that value in polar regions during that hemisphere's winter. Over the continental United States, then, the average tropopause height in winter is around 35,000 feet, and it lowers to around 18,000 feet in winter (Fig. 2-2). This is important even to pilots of light airplanes because the turbulence effect of the jet stream can extend to 20,000 feet beneath the tropopause height, enough to impact turbocharged and turboprop airplanes in summer and even normally aspirated aircraft in the winter.

GLOBAL WIND PATTERNS

Warm air rises and cold air descends, and the spinning of the earth induces a force called the *Coriolis effect*. These facts are the basis for global wind-flow patterns.

Air over the equator is heated and rises, descending over the polar regions. A "boiling" pattern begins, with a descent of air over the mid-latitude regions, creating a constant circulation that results in the general flow of the wind. As the air circulates, the earth beneath rotates. This creates an apparent bending of the air flow relative to the surface. From the equator to around 30 degrees north latitude, about the southern edge of the continental United States, the apparent wind flow is typically southwest to northeast. From about 60 degrees north to the pole, in descending, polar air, the pattern is reversed; high-latitude pilots enjoy patterns opposite their counterparts to the south. The continental United States is in a calm area between the two regions (Fig. 2-3);

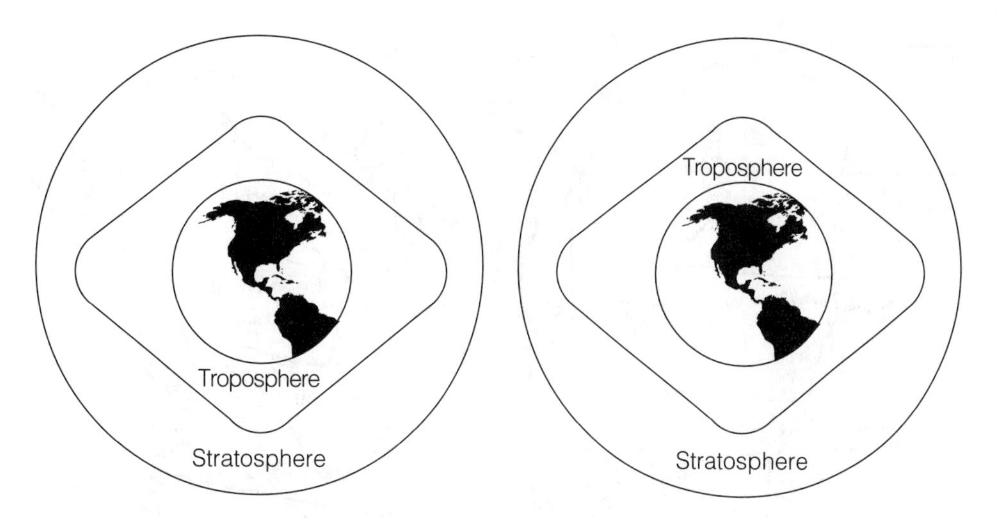

Fig. 2-2. *Tropopause height during the northern hemisphere winter (left) and summer (right).*

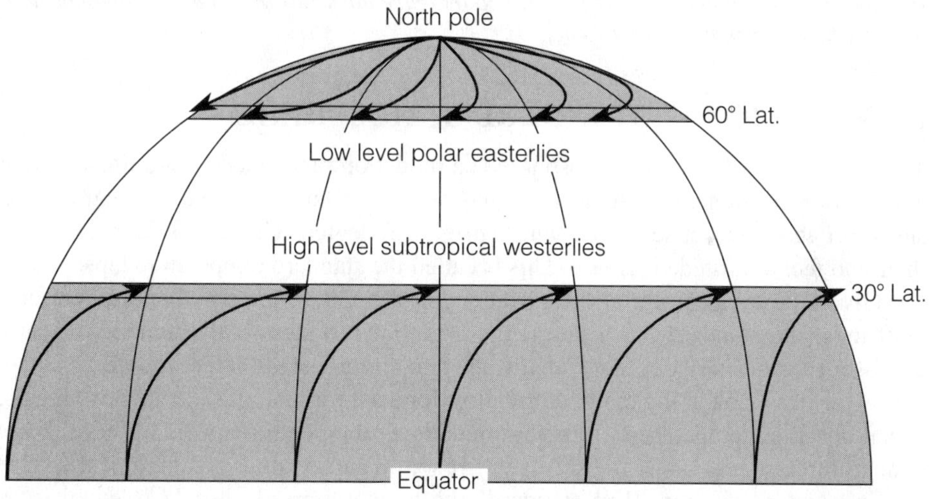

Fig. 2-3. *Northern hemisphere wind patterns.* Advisory Circular 00-6A

however, the influence of the southern area is greatest, and more warm, moist air is available in the south, so most of our weather travels from southwest to northeast.

The Coriolis effect creates the spiraling motion of water down a drain. In the same way, air circulates clockwise around and away from high-pressure areas and counterclockwise into areas of low pressure. You can plan a long-distance flight around pressure areas, then, to optimize tailwinds or reduce headwinds (Fig. 2-4). We'll cover more about the effect of these flow patterns in chapter 4.

Fig. 2-4. *Flying around pressure patterns to increase tailwinds or reduce headwinds might actually be faster than a direct route.*

INTERNATIONAL STANDARD ATMOSPHERE

The International Standard Atmosphere is a model of atmospheric conditions used as a basis for measurement. ISA, as it's called, assumes a temperature of 15 degrees Celsius (59 Fahrenheit) at sea level, with a drop of 2 degrees Celsius (3 Fahrenheit) per thousand feet of altitude increase. This is called the standard temperature lapse rate.

Given the temperature at the surface, you can determine whether temperatures aloft meet the standard lapse rate (Fig. 2-5). For instance, if the temperature at a 2000-foot airport is 10 degrees Celsius, the temperature at 9000 feet should be –4 degrees Celsius. Look at the temperature aloft forecast for that altitude, or compare the actual observed temperature from your outside air temperature gauge in flight. Is the temperature actually –4 degrees Celsius? If so, the lapse rate is standard. If the lapse rate is standard or more than standard (the temperature at altitude is equal to or colder than expected), then the air is stable. Anticipate a smooth ride and little chance of thunderstorm development. If, however, the air is warmer than the standard lapse rate suggests, then the air is unstable; plan on turbulence and keep an eye out for convective activity. You can also predict freezing levels using the standard lapse rate model.

The International Standard Atmosphere also assumes an air pressure of 29.92 inches of mercury at sea level (1013.2 millibars), with a drop of 1 inch per 1000 feet in the lower levels of the atmosphere. Although moisture affects the temperature lapse rate, humidity isn't figured into the ISA.

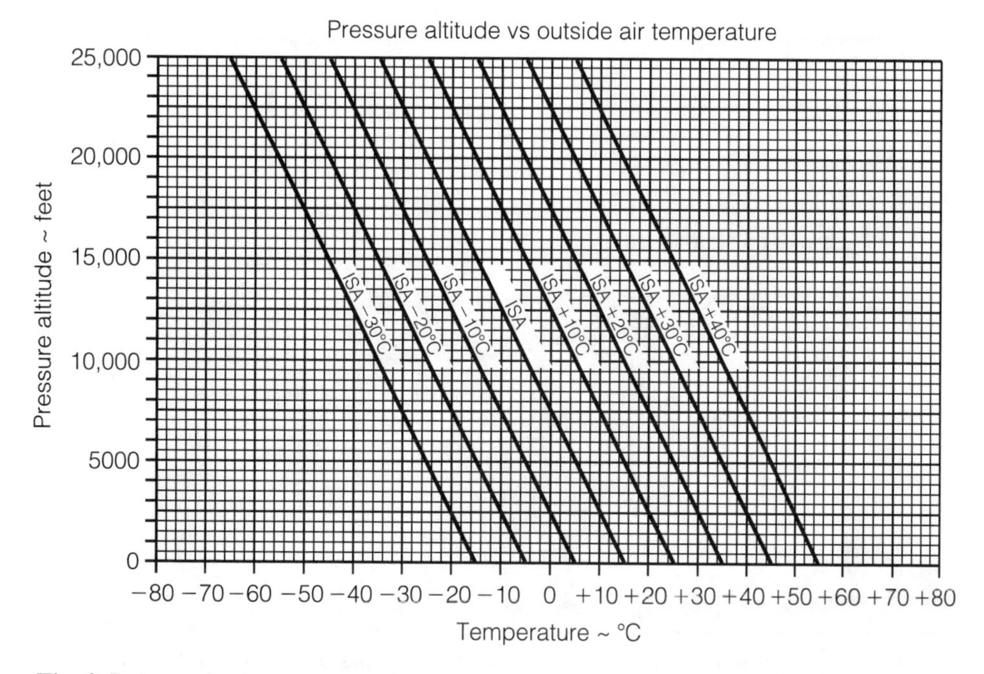

Fig. 2-5. *A standard temperature-lapse-rate chart.*

NONSTANDARD LAPSE RATES

ISA is an idealized measuring tool, but the atmosphere rarely conforms to this ideal. Instead, we sometimes have nonstandard lapse rates which, as we've already seen, can affect the weather.

The pressure lapse rate might be nonstandard if "bubbles" or "pockets" of hot or cold air are pushed aloft. Pockets of lifted warm air are sometimes called *lows aloft* or *upper-level disturbances*, while areas of cold air are designated *highs aloft*. Because warm air is less dense than cooler air, and cool air flows to fill this vacuum, lows aloft create a lifting tendency in the air beneath them, exaggerating any other lifting mechanisms to create rain showers or storms. Lows aloft create instability. Conversely, a high aloft is an area of descending air and tends to cancel out any underlying source of adverse weather. Highs aloft add stability to the atmosphere. Use this knowledge to create a three-dimensional picture of the atmosphere when you evaluate the likelihood of adverse weather (Fig. 2-6).

Temperature, too, often exists in nonstandard lapse rates. An *inversion* is an area where warm air overlies cooler air. Condensation and turbulence commonly occur at the lower boundary of an inversion; expect reduced visibility or fog.

Inversions can be formed by the passage of a weather front (chapter 4). They are also common over land in the morning after a cool, clear night with light winds. In the last part of the night, the rapid cooling of the earth's surface transfers to the lowest

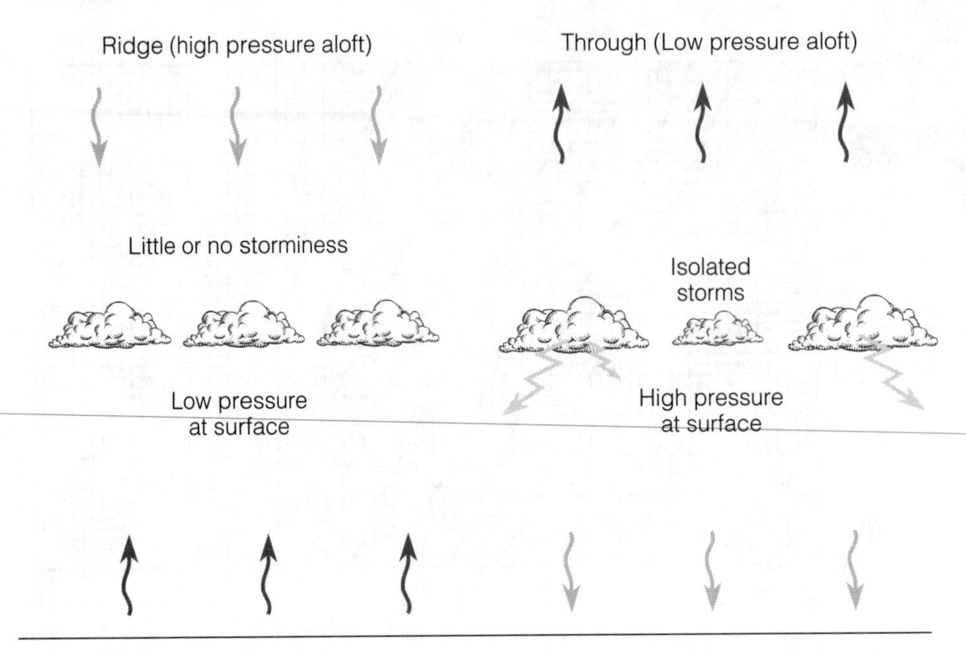

Fig. 2-6. *A three-dimensional weather picture allows a more accurate prediction.*

layer of air, cooling it more than the warmer air rising above it. This forms the typical early-morning "haze layer" you see in the summer, when a climb through the first thousand feet or so of the air produces a dramatic increase in visibility. You might feel a little turbulence as you climb through the top of this layer also. When solar radiation again begins to warm the earth, a new layer of warm air forms near the ground, pushing this cold layer upward; the point of contact between dissimilar air masses rises, taking the reduced visibility with it (Fig. 2-7). This is what causes fog or this haze layer to "lift" before the cold air aloft itself is warmed, allowing the condensation to return to vapor state and clearing the air.

Here are some rules of thumb about the makeup of the atmosphere:

- The jet stream, a layer of turbulent, high-speed air at the tropopause, can affect airplanes as low as 15,000 feet in summer and at sea level in winter.
- You can use the surface temperature and the temperature forecast or observed at altitude to determine the air's stability. Do this by comparing the temperatures to the standard lapse rate; stable air tends to improve conditions, while unstable air worsens them.
- Highs aloft add stability to the atmosphere, while lows aloft (sometimes called upper-level disturbances) add instability.

- Knowing the standard lapse rate also helps you to forecast the freezing level.
- Inversions almost always cause reduced visibility or fog, and are common in the morning after a cool, clear night with light winds. Conditions will improve as sunlight warms the surface.

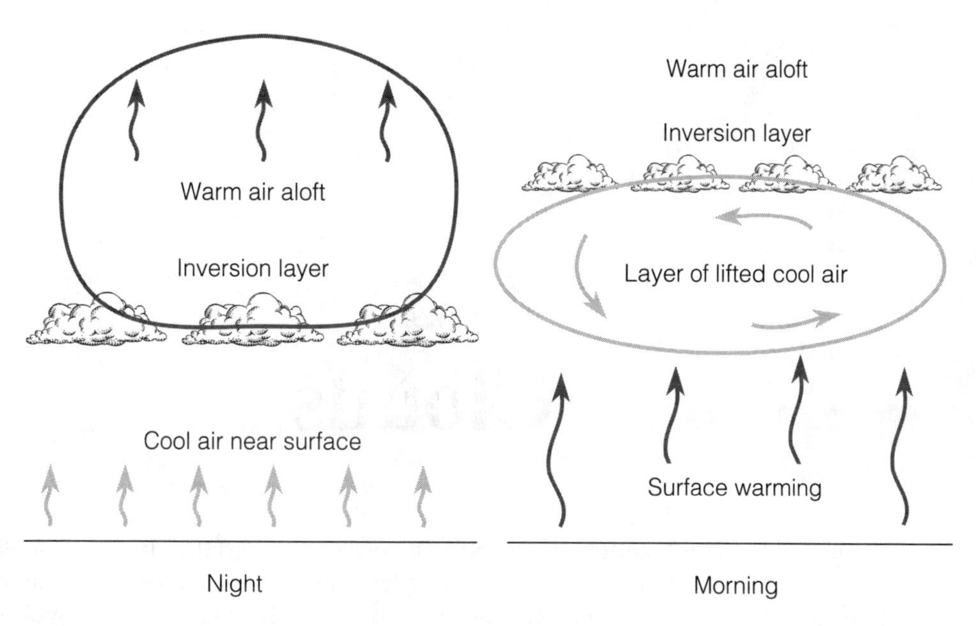

Fig. 2-7. *An inversion forms on cool, clear nights (left) and lifts with the morning's heating (right).*

3
Clouds

CLOUDS ARE THE VISIBLE SIGNPOSTS OF MOISTURE IN THE AIR. THEY APPEAR whenever condensation occurs, that is, whenever the air is saturated at that altitude. Generally, the more moisture available, the lower the cloud base and the thicker the cloud layer. Clouds might or might not signal precipitation.

Look in a meteorology textbook and you'll find literally dozens of types of clouds. From a pilot's standpoint, however, you really only need to know two things about a cloud to determine the hazards it might represent. You need to know whether it is of the cumulus or stratus variety, and you need to know its approximate altitude.

Clouds come in two varieties: cumulus, the puffy, bumpy clouds shaped by turbulence and lifting in the atmosphere; and stratus, which form in smoother skies (Fig. 3-1). Each variety has its own associated hazards.

Cumulus clouds indicate a bumpy ride. If the air were smooth, a cumulus cloud would not form. The turbulence isn't limited to the cloud itself; the visible cloud exists like whitecaps on waves of the air, so expect a bumpy ride below and around the cumulus clouds as well. It usually smooths out above the tops of cumulus clouds.

If enough moisture is present, cumulus formations can cause precipitation. Because of the isolated nature of this type of cloud, expect the rain or snow to be showery, or confined to small areas, and relatively short in duration. Often there is lightning and thunder along with this rain or snow; we'll look at thunderstorms in detail in a later chapter.

Cumulus clouds

Unstable air
Turbulence
Showery precipitation
Clear ice

Stratus clouds

Unstable air
Smooth ride
Widespread precipitation
Rime ice

Fig. 3-1. *Characteristics of cumulus clouds (left) and stratus clouds (right).*

Airframe ice can form quickly in a cumulus cloud. Because of the updrafts that form this cloud type, relatively large droplets of water can be held aloft. Run into these big drops with a cold airplane, and they'll spread out to freeze as clear ice, difficult to remove. Ice accumulation is usually heaviest at the tops of cumulus clouds.

Stratus clouds, on the other hand, warn of different hazards. Formed in smooth air, stratus clouds indicate a comfortable ride. Possessing a huge amount of moisture, the stratus formation can bring days-long rains or snows to a single location and reduce visibility at the surface for days. A general rule of thumb is that if significant rain or snow is hitting the surface, the cloud layer is at least 4000 feet thick. This can help you flight-plan around possible icing conditions, as we'll later see.

With little lifting action in the air, stratus clouds suspend only small droplets of water; hence, any icing in stratus formations is usually of the rime variety, more easily removed than clear ice. The rate of ice accumulation is less than in a cumulus cloud, but it can still be dangerous if flown in long enough.

You can tell a lot, then, about the hazards facing a flight simply by identifying the variety of a cloud. To carry this further, you can anticipate from your weather briefing the types of clouds you'll encounter en route, and you can use this forecast to verify or refute the official weather briefing. We'll cover this in detail later. To get the "big picture," however, you'll need one more bit of information: the height, or altitude, of the cloud.

Clouds are divided into four groups, or "families," identified by their height. Different families warn of different flight hazards. The first family, the high clouds, range from about 15,000 feet to 30,000 feet and higher (Fig. 3-2). At these altitudes, visible moisture usually consists of ice crystals. Hence, these "cirro" type clouds bring little danger of airframe ice, since ice crystals will simply bounce off an airplane. There might be turbulence in high clouds if they're of the cumulus type.

If you do fly through high clouds, the most likely hazard is precipitation static. When you fly through precipitation, an electric charge builds up on the airframe. Many airplanes have static wicks attached to the wings and tail to help dissipate this electric-

Fig. 3-2. *High clouds.*

ity. Precipitation static can interfere with radio communication and navigation; lorans are especially susceptible to "p-static" outages.

Because most airplanes can't fly high enough to penetrate the high clouds, most pilots are concerned with the weather beneath them. High clouds block the inflow of solar radiation, so there's less convective turbulence and little chance of thunderstorms beneath them. Expect a smooth, pleasant ride; this is the sort of day for a relaxing flight or for introducing a friend to aviation. High clouds can signal the approach of a warm front or the passage of a weak trough, which we'll see in the next chapter.

The second family of clouds, the intermediate or mid-level clouds, can herald any of the hazards to aviation. Designated the "alto" type, these range from around 5000 feet to 15,000 feet or more (Fig. 3-3). Altocumulus clouds sometimes bunch into a near-overcast, warning of turbulence, showery precipitation, and, if the temperature is right, large amounts of clear ice. Altostratus clouds bring day-long rain or snow, so expect poor flying weather.

The third family of clouds, the low clouds, hover below the 5000-foot point (Fig. 3-4). Cumulus clouds in this range usually contain little moisture, so anticipate little precipitation or ice, although the ride might be bumpy. Stratus clouds this low might be thick and moist, soaking the surface beneath and containing large amounts of rime ice in winter.

The fourth and final family of clouds are those of extensive vertical development. As the name implies, these clouds are often 30,000 feet or more in height, with bases

Fig. 3-3. *Intermediate, or mid-level clouds.*

Fig. 3-4. *Low clouds.*

near the surface but averaging about 5000 feet above ground (Fig. 3-5). Clouds require significant lifting and turbulence to reach these heights; therefore, they are always of the cumulus variety. These are the massive thunderheads that bring locally heavy rains, hail, lightning, and thunder; because of their height at the top they can contain heavy, clear ice even in summer. We'll discuss thunderstorms and ice in later chapters, but the "short course" on these clouds is to avoid them.

Fig. 3-5. *Clouds of extensive vertical development.*

You can tell a lot about the hazards present simply by identifying the variety (cumulus or stratus) and the altitude of clouds. As we'll see in the next chapter, you can anticipate the types and locations of clouds you'll encounter when you receive a preflight weather briefing; to evaluate the true weather picture en route, compare your expectations to what you actually see. Clouds are the visible signposts of weather.

Important points:

- Cumulus clouds indicate instability, turbulence, showery precipitation, and the possibility of heavy, clear ice.
- Stratus clouds indicate stability, a smooth ride, long periods of precipitation, possible low visibility near the surface, and the possibility of rime ice.
- The altitude of a cloud helps identify potential aviation weather hazards.

- If significant precipitation is falling at the surface, the producing cloud layer is usually at least 4000 feet thick.
- Types of cloud formations are predictable for given weather patterns. You can anticipate the type and location of clouds you'll encounter on a trip, and you can use the types and locations you actually observe to verify or refute the forecast.

4
Fronts

MOST OBSERVED WEATHER, AS WE'VE DISCUSSED, OCCURS AT THE BOUNDARY between dissimilar air masses. These boundaries, or battle lines between properties of temperature and moisture, are called *fronts*. Fronts are usually marked by a change in temperature, humidity, and pressure. They are always identified by a change in the wind flow. Different types of fronts have predictable and different types of hazards. How do fronts form?

In the northern hemisphere, typically, air masses are warm and moist in the south and colder and drier in the north. Before pressure patterns, and therefore fronts, begin to develop, a stationary front will form between these two air masses (Fig. 4-1). A *stationary front* is a boundary where the forces blowing the system in one direction are about equalled by wind flow from the other direction; we'll talk about the hazards of a stationary front later in this chapter.

This stationary system is not stable, and soon the warmer air to the south begins to form a low pressure area. A counterclockwise wind flow begins, pulling colder, usually drier air down along the west side of the low (Fig. 4-2). A *cold front* forms when cold air is advancing against warm air; cold fronts tend to form along the west side of a low and circulate around to the south.

Meanwhile, the low pressure circulation forces warmer, southern air up around the east side of the center (Fig. 4-3). A *warm front* forms when warm air overrides colder

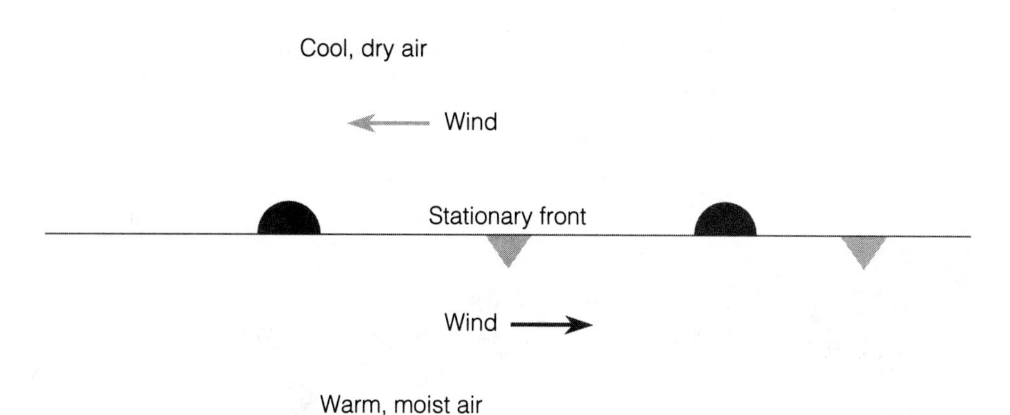

Fig. 4-1. *A stationary front.*

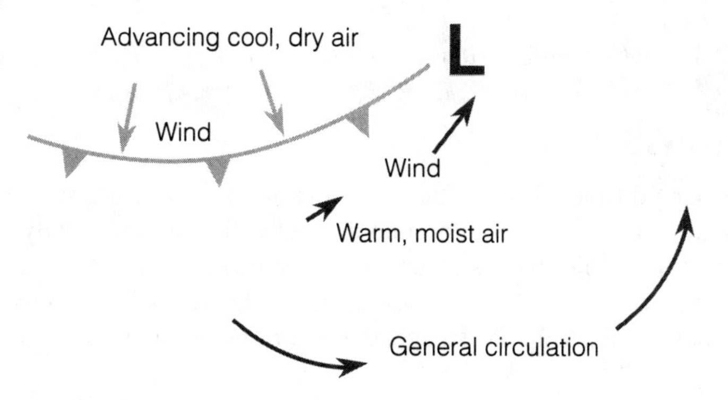

Fig. 4-2. *The beginnings of a cold front.*

air; warm fronts form in the south or southeast quadrant of a low and rotate around to the east or northeast.

It's important to note that weather systems do not merely form and blow around the globe; they develop and moderate, or change, as they come into contact with warmer or cooler surfaces. The intense cold front to the west of a low, then, gradually warms with contact with the warmer air and surface on its way around the low; the temperature contrast diminishes, and the system gradually runs out of energy. It might even turn into a warm front as counterclockwise air flow pulls it around to the east. The warm front working its way up the east side of the pressure system cools with contact with the northern regions and dies out before pointing much beyond an east-west line.

There are other types of fronts a pilot needs to be concerned with, but for now we'll concentrate on the two most common and obvious types of fronts; the cold front and the warm front.

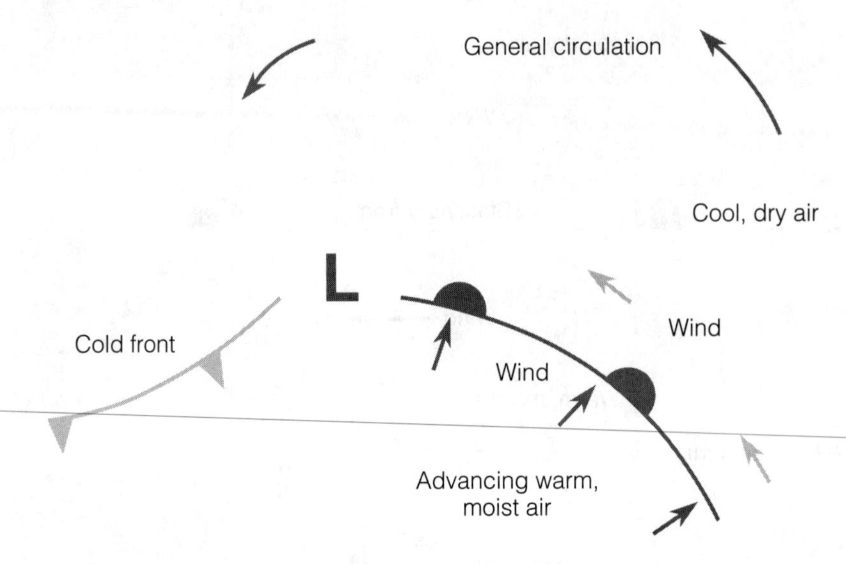

Fig. 4-3. *The beginnings of a warm front.*

COLD FRONTS

As we've seen, cold fronts form on the west side of a low pressure system and move around the southern quadrant, coming into contact with warmer, usually moister, air. On a weather chart, a cold front is depicted as a line with triangle symbols indicating the direction of flow (Fig. 4-4). In cross-section, the classic cold front looks like an inverted bowl moving along the landscape, dense air hugging the surface and wedging

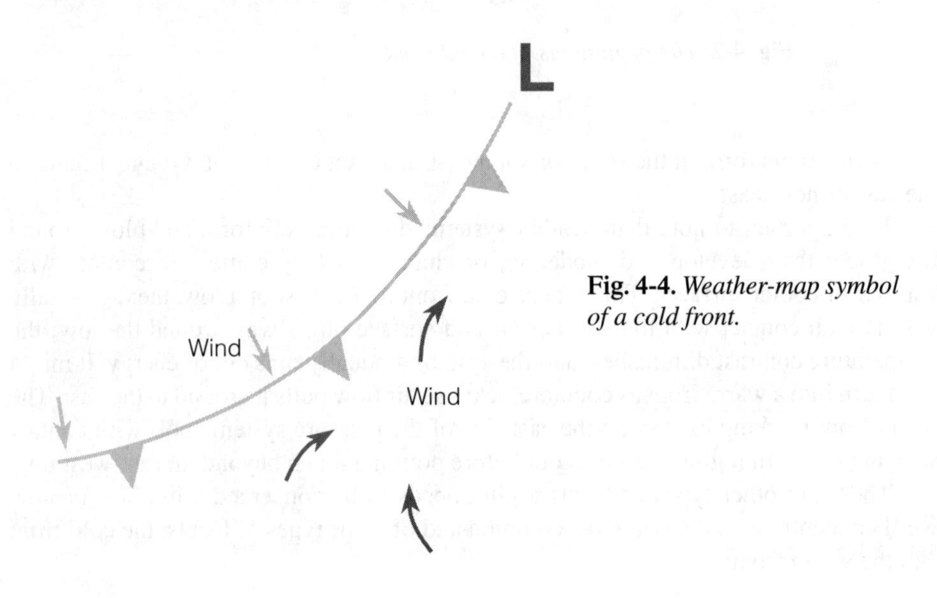

Fig. 4-4. *Weather-map symbol of a cold front.*

lighter, warmer air aloft (Fig. 4-5). This lifting action tends to create turbulence and thunderstorms or snow squalls, marked by cumulus clouds and all their associated hazards to aviation. The line of adverse weather, however, tends to be quite narrow, maybe even only a few miles. So if the clouds are not penetrable by pilots, at least the system will blow over quickly.

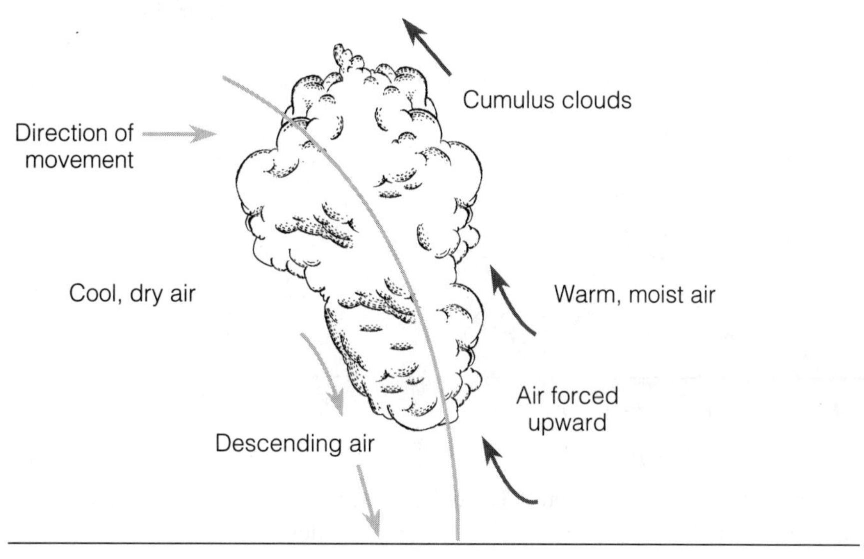

Fig. 4-5. *Cross-section of a typical cold front.*

Cold fronts average 25 to 50 knots ground speed. Look for a big temperature difference ahead of versus behind the front; if the temperature drop exceeds 20 degrees, expect a lot of adverse weather. Strong cold fronts, identified by ground speeds in excess of 35 knots, are preceded by a strong southerly wind flow, air being pushed out of the way by the advancing cold air. The closer to the pressure center, the stronger the winds will be. With very strong cold fronts, shock waves in the air might form like the ripples ahead of a boat on calm water; these shock waves, called *squall lines*, act like miniature cold fronts themselves and can contain some of the most severe weather, 50 to 300 miles ahead of the front. If flying toward a cold front from the east, look for mid-level cumulus clouds 100 miles or more ahead of the front, and expect the winds aloft to increase out of the southwest until crossing the boundary. These cumulus clouds might become extensive and develop into thunderstorms; when crossing the front proper, you'll encounter more clouds of extensive vertical development, storminess, pressure and wind changes, and turbulence to 10,000 feet or more (Fig. 4-6).

If the temperature change crossing the front is less dramatic, say, less than 20 degrees, or if the front's speed is less than 35 knots, the pattern changes. Less contrast means less wind, and the steep frontal slope normally associated with cold fronts be-

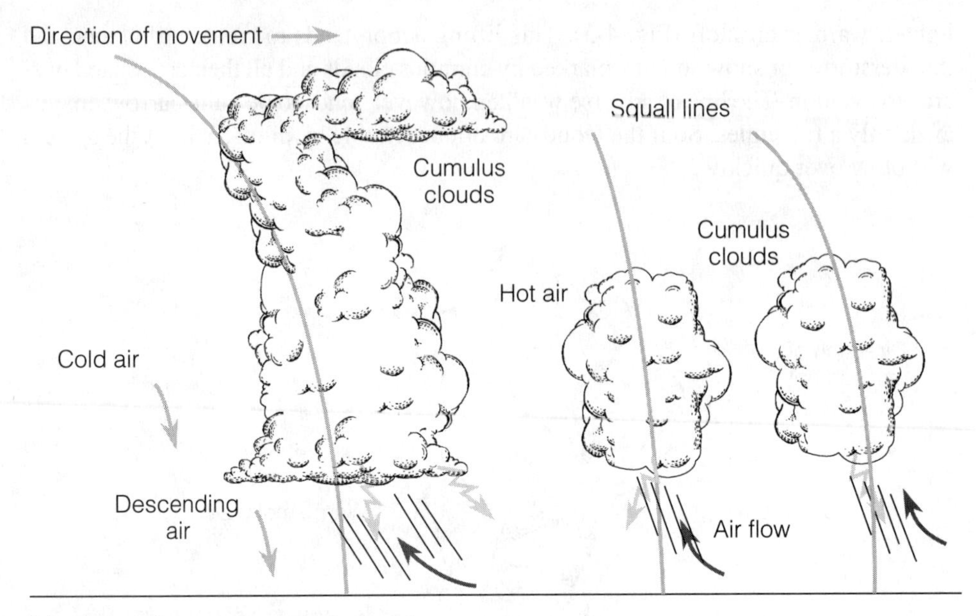

Fig. 4-6. *A fast-moving cold front.*

comes a more gradual boundary. Expect this sort of pattern as the front revolves around the south side of the pressure system and moderates.

With this gentler slope, the point of contact between air masses is extended, creating a wide area of cloudiness. With less contrast, the air is more stable, so stratus-type clouds will form, but there might be enough lifting action to create smaller cumulus clouds and thunderstorms embedded in the murk (Fig. 4-7). Hence, a slower cold front is less stormy than its faster counterpart but might actually be more dangerous to pilots. Expect low- to mid-level stratus clouds approaching the front, and watch for signs of turbulence or cumulus tops sticking out of the layer to warn of invisible yet hazardous storms. Mixed icing is also possible if temperatures are right, as you'll encounter both cumulus and stratus clouds. Since these fronts' ground speeds are low, expect the hazards to exist across a given area for some time, and be prepared to reroute or wait it out unless you and your airplane are equipped to detect embedded storms aloft.

Whether fast or cold, steep or gradual, the passage of a cold front brings a shift of winds from the northwest, colder air, and a rapid rise in air pressure. The sky is usually clear and bumpy, but it might become cloudy or foggy if the frontal weather deposited a lot of moisture that's ready for condensation in colder air.

WARM FRONTS

Warm air rises, so a warm front has a broader cross-section than a cold front. The advancing, warm air will slide over the descending, cooler air, usually creating an area of extensive cloudiness several hundred miles wide (Fig. 4-8). As a warm front ap-

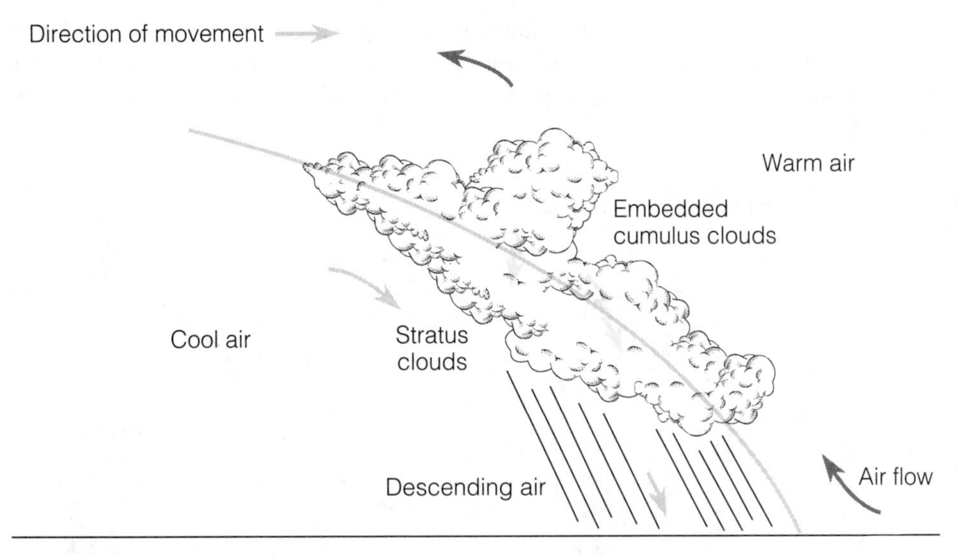

Fig. 4-7. *A slow-moving cold front.*

proaches, or as you fly toward the front from the cold side, you'll typically see high-level, cirrostratus clouds as much as 700 nautical miles ahead of the front. Flying weather will be good beneath this veil of clouds, which will gradually slope downward when progressing toward the front. Eventually, as the clouds drop to the mid-levels, precipitation usually begins—a slow, drawn-out rain or snow that can last days at a time. Approaching the front at the surface, cloud bases and visibilities drop, clearing slowly after the front passes.

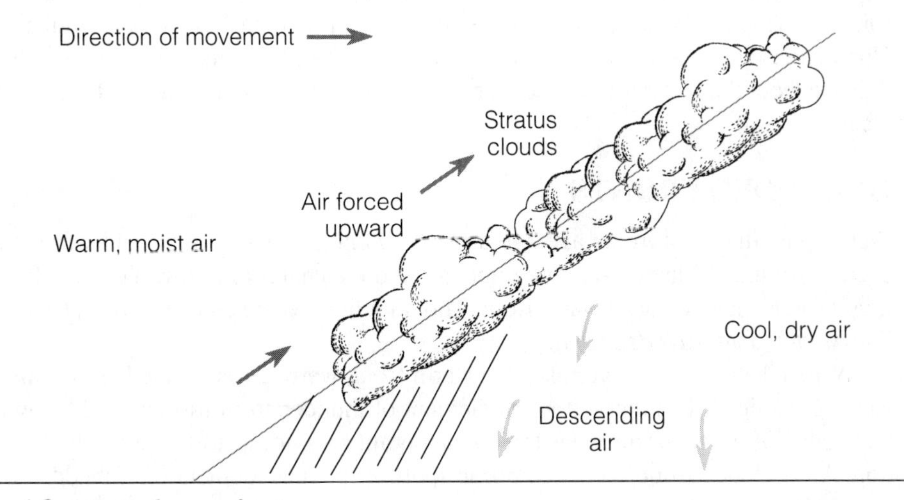

Fig. 4-8. *A typical warm front.*

Warm fronts average around 15 knots ground speed. Faster-moving warm fronts tend to form a steeper frontal slope, ramming against the cold, denser air; this steep frontal slope looks much like a slow-moving cold front and creates similar hazards (Fig. 4-9).

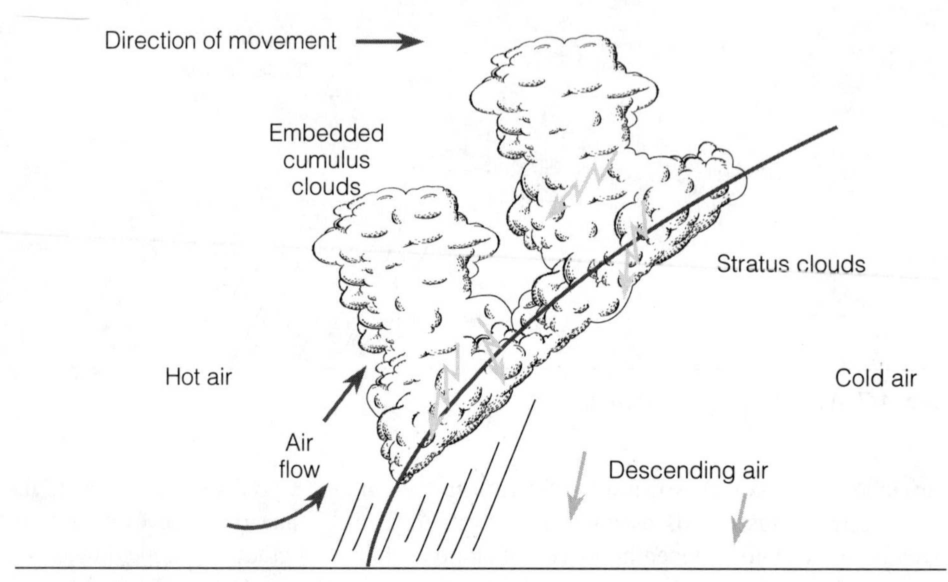

Fig. 4-9. *A fast-moving warm front.*

Flying towards a warm front, in addition to the lowering cloud bases, you'll note a falling barometer; near a front, be sure to get updated altimeter settings to avoid errors. Winds might be easterly or southeasterly, as they're pushed away from the front. Upon entering the warm sector, you'll notice a shift of the winds from the south, turbulence from the instability of the warmer air, and a slow, sometimes only temporary, clearing.

OCCLUDED FRONTS

We've seen that cold fronts average around 30 knots ground speed, and warm fronts average around 15 knots. As they rotate around the center of a low, then, a cold front might catch and overtake a warm front. This situation, where cold and warm fronts collide, is called an *occluded front*.

When the cold front overtakes the warm front, warm air is wedged aloft. This creates a great deal of lifting action, an element of thunderstorm formation. Meanwhile, the wedge of warm air provides the contact point for widespread stratus cloud formation. An occluded front, then, is a prime spot for the development of embedded thunderstorms (Fig. 4-10).

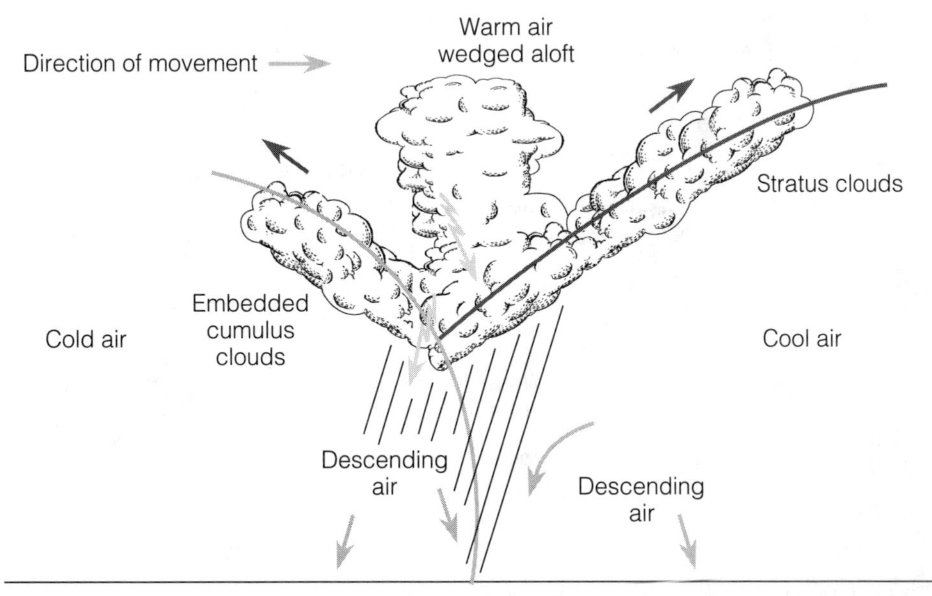

Fig. 4-10. *Cross-section of an occluded front.*

As a rule, flying weather is very poor along "occlusions," which contain none of the "good" points and all of the hazards of both warm and cold fronts. Embedded thunderstorms, wind shear, and widespread mixed icing are all common in these formations. The most severe weather occurs at the point of contact between the fronts, as this is where the greatest contrasts occur; this point of contact will slide southeasterly along the fronts as the counterclockwise rotation continues (Fig. 4-11). If at all possible, time your flight or pick a route to avoid all areas of an occlusion, but, if you must penetrate the system, avoid the point of contact.

The good news is that occluded fronts are relatively short-lived. Once the warm air is replaced at the surface with cold, the convection necessary to add moisture to the front is lost, and the system quickly wears itself out. Occluded fronts tend to die out in a day or two, the last gasp of a storm system before contrasts fade and the low dies away. This leaves a stationary front south of the original low, a building block for yet another frontal zone to develop in the future (Fig. 4-12).

There is a definite life cycle to a frontal system, then, from a cold front to an occlusion and a warm front; each type of front has its own set of hazards (Fig. 4-13).

OTHER TYPES OF FRONTS

Cold, warm, and occluded fronts are the classic battlegrounds that create adverse flying weather, but there are other types of temperature/moisture contrasts that affect pilots as well. Now we'll look at some other types of fronts and the hazards they present.

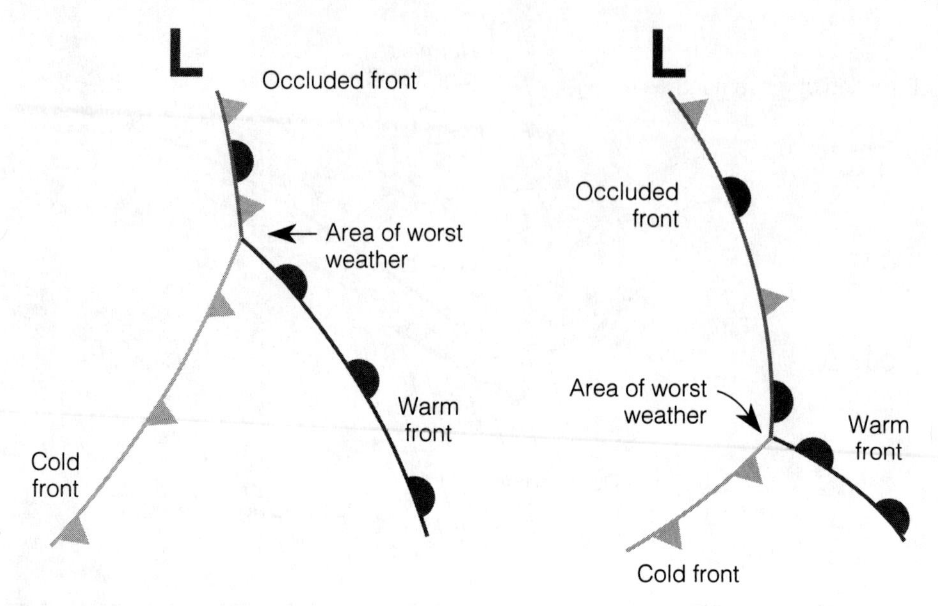

Fig. 4-11. *Development of an occluded front.*

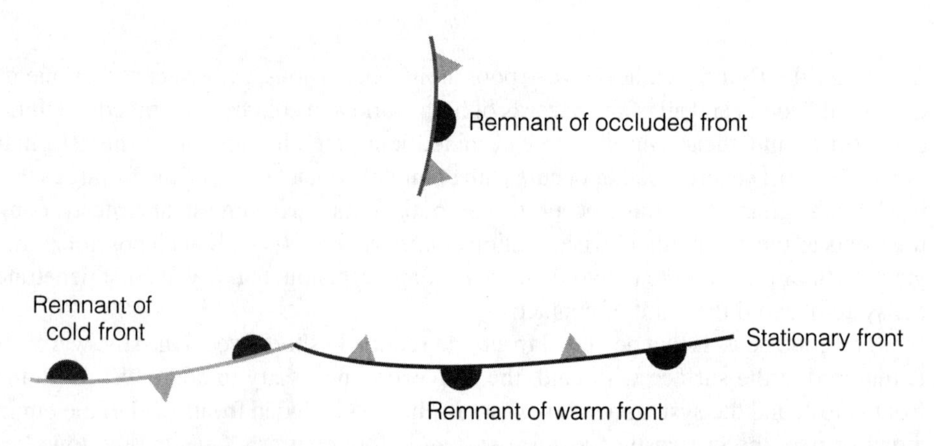

Fig. 4-12. *"Death" of an occluded front.*

Stationary fronts

As discussed earlier, stationary fronts form when the wind forces on one side of the front roughly equal the wind forces on the other side of the front. Normally, this would indicate small pressure, temperature or moisture contrasts, meaning little adverse weather. There are some situations, however, where the wind forces might be in balance even though a strong contrast exists.

One region that sees a lot of stationary fronts is the American southeast. Cold fronts roar across the plains states and into the southeast, only to butt against the strong

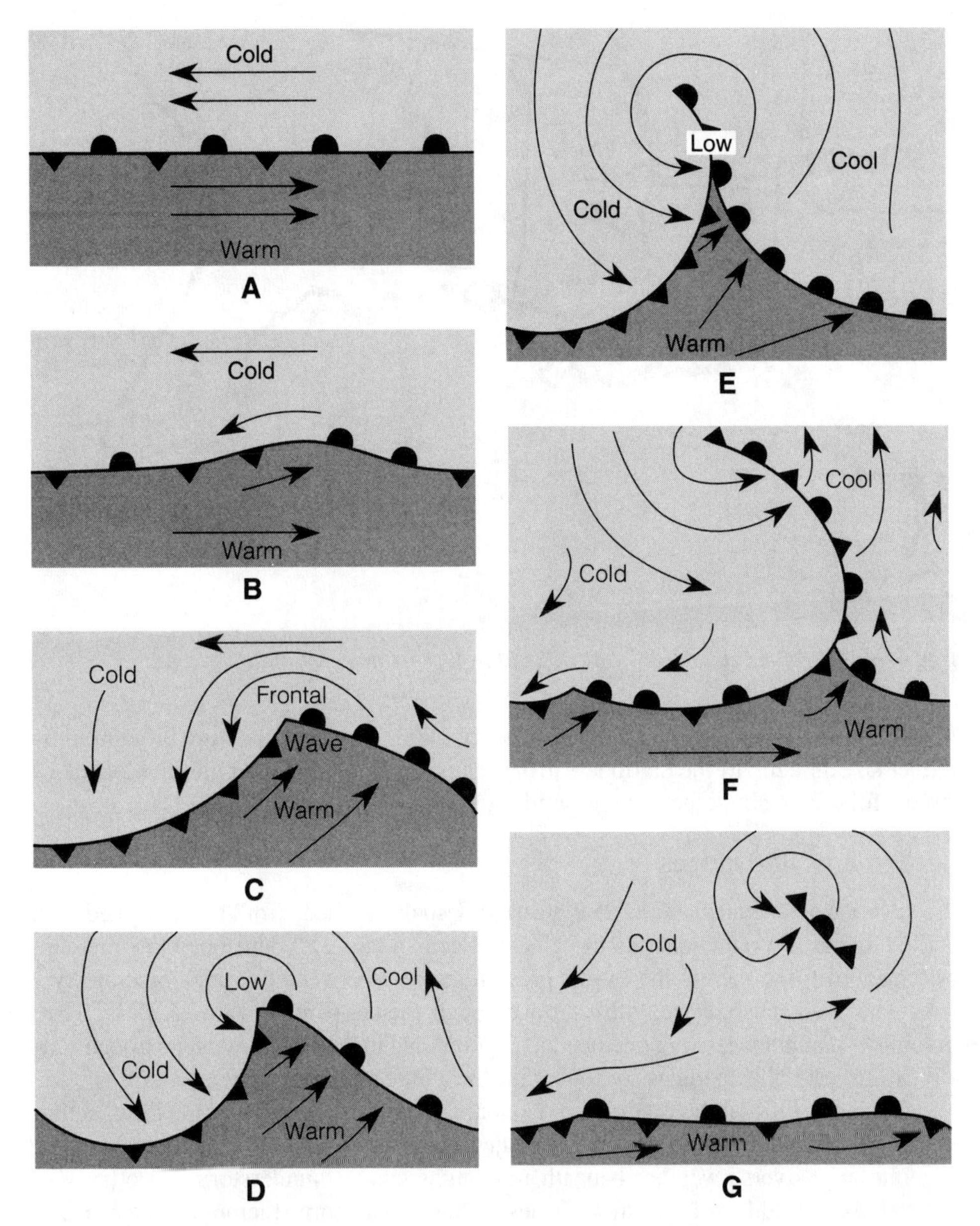

Fig. 4-13. *Life cycle of a frontal system (Advisory Circular 00-6A).*

influence of the semipermanent Bermuda high. The clockwise-spinning, dense air stops the advancing cold front in its tracks, and places like Atlanta settle in for a three- or four-day rain (Fig. 4-14).

Anytime the jet stream flows along and over a front, it might also hold that front in place, creating an extended period of heavy precipitation like that in the Upper Mis-

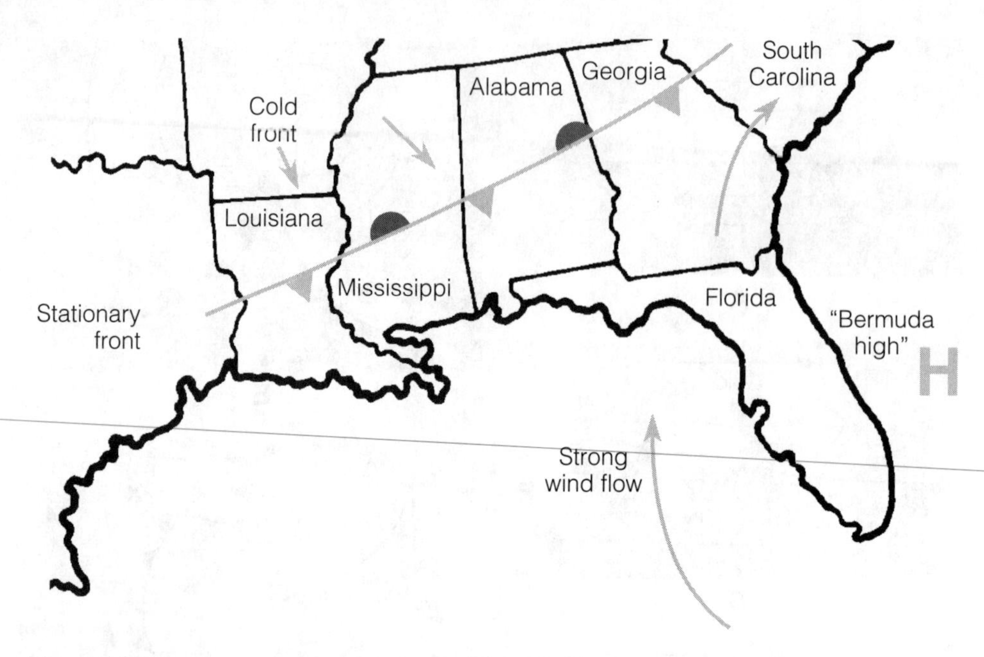

Fig. 4-14. *The "Bermuda High" halts a cold front, creating a stationary front.*

sissippi and Missouri valleys in the summer of 1993. Any new moisture brought in by the jet stream adds to the stationary front, causing "training," or repetitive bands of storms following about the same ground track (Fig. 4-15).

Troughs and ridges

A *trough* is an elongated area of low pressure. Usually spelled "trof" and indicated by a dashed line on a weather chart, a trough is basically a less-well-developed low-pressure system, sometimes merely the lowest pressure evident between two high-pressure systems. The weather associated with a surface trough, then, is similar to a weak warm front; in summer it might mean rain or embedded storms, and in the cold of winter a trough usually brings just light cloudiness or fog, or maybe a little precipitation (Fig. 4-16).

A trough aloft, on the other hand, is an area of warm air at the tropopause, acting as a giant vacuum cleaner as colder air rushes upward to fill the void. This tends to exaggerate any adverse weather beneath it; troughs create thunderstorms in otherwise clear skies and add to the destructiveness when other storm factors are at work. A trough aloft, sometimes called a "disturbance," is indicated by a dip in the jet stream toward the south; if you see a surface weather chart depicting a low pressure system with no illustrated fronts, often this refers to an upper-level disturbance.

Similarly, a *ridge* is an elongated area of high pressure, pushing away adverse weather. A ridge aloft consists of dense, descending air, weakening or even canceling completely any lifting action below to effectively dampen out thunderstorms or other hazards. Ridges aloft are indicated by a northward bend in the jet stream path (Fig. 4-17).

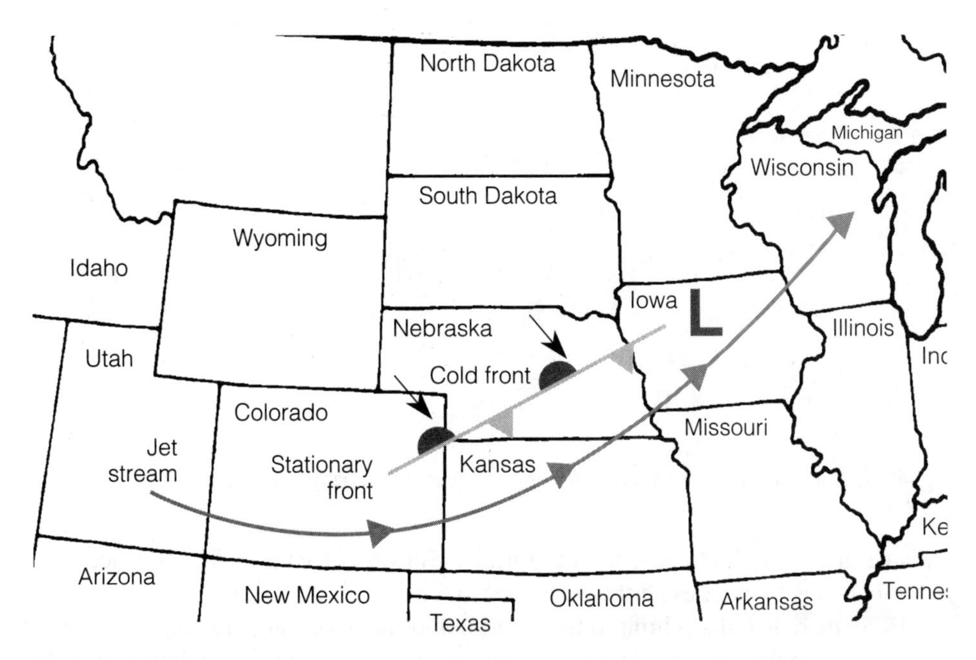

Fig. 4-15. *The jet stream blocks a cold front, creating a stationary front.*

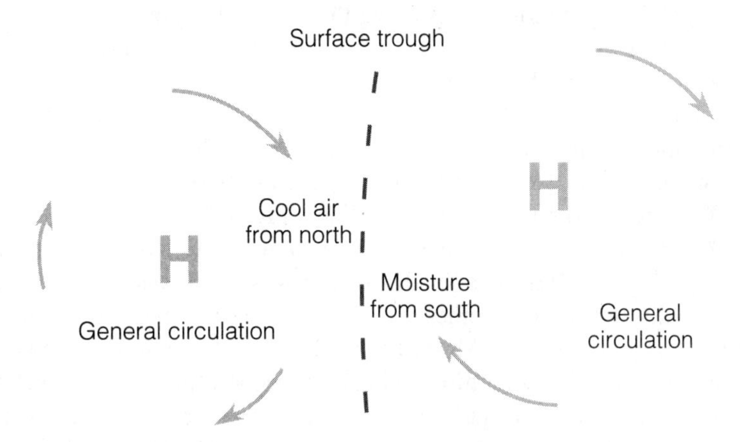

Fig. 4-16. *A trough between high-pressure areas.*

Dry lines

Ask most pilots what a dry line is, and they're at a loss. Dry lines aren't often mentioned in aviation weather forecasts, but they're a source of some of the most severe convective activity in the central part of the United States.

Assuming that a big high pressure system reigns in the central United States, warm, moist air from the Gulf of Mexico expands overnight to influence areas inland. In the east-

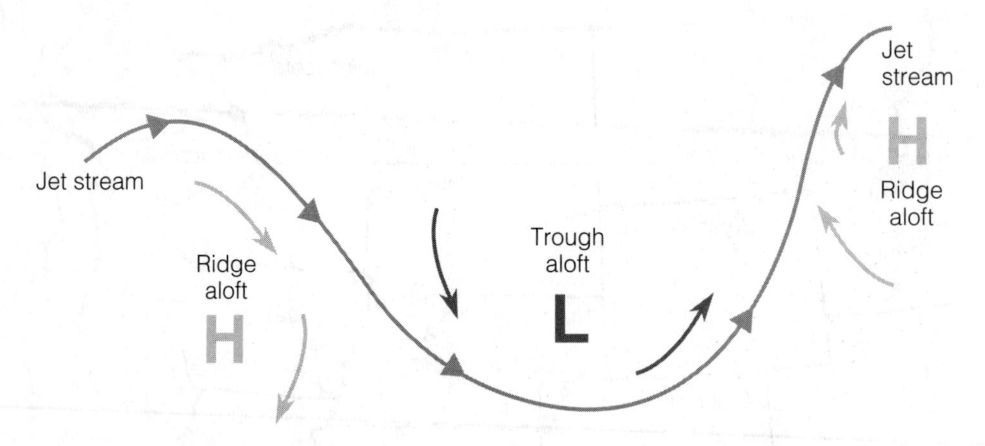

Fig. 4-17. *Bends in the jet stream indicate the locations of ridges and troughs.*

ern states there is little effect, since those areas are always moist from the influence of the Atlantic. In the dry prairies northwest of the Gulf, however, it's a different story.

Here, there is a sharp limit to the expansion of the moist air, creating two distinct air masses, a moist one and a dry one. Temperatures might be very similar on one side versus the other, so no classic pressure pattern forms, but moist air comes into contact with drier air, and evaporation begins along the border. Evaporation is a cooling process, so the air to the west of the *dry line* begins a gentle flow toward the east; the sharp demarcation of the dry line begins to look very similar to a fast-moving cold front, although ground speed is small. Consequently, strong storms might form and move slowly across the ground, blown by the general easterly flow; the two most powerful tornadoes on record, the Hesston, Kansas, tornado of 1990 and the Andover, Kansas, tornado in 1991, were both the result of dry lines (Fig. 4-18).

Dry lines are so common in Texas that meteorologists have a special name for them: the *Marfa front*. Marfa is a small town in western Texas; dry lines form near there almost daily unless some other weather pattern is blowing through. A line of storms might form on a roughly Marfa-Lubbock line in the morning, slowly blowing east to a roughly Dallas-San Antonio line in late afternoon (Fig. 4-19). Watch the satellite or radar time-lapsed pictures of Texas on a high-pressure, summer day and you'll likely see the Marfa front; if planning a flight to Texas, time your arrival to miss the approach of this front.

"LOW PRESSURE=BAD WEATHER"

This is an axiom of the purveyors of home barometers. As I write this I'm looking at my own home barometer, where the needle points at the word "Fair" because the air pressure is about 30.10 inches of mercury. A few inches to the left of my barometer, freezing drizzle taps my window, certainly not "fair" weather, especially for a pilot. Why the discrepancy?

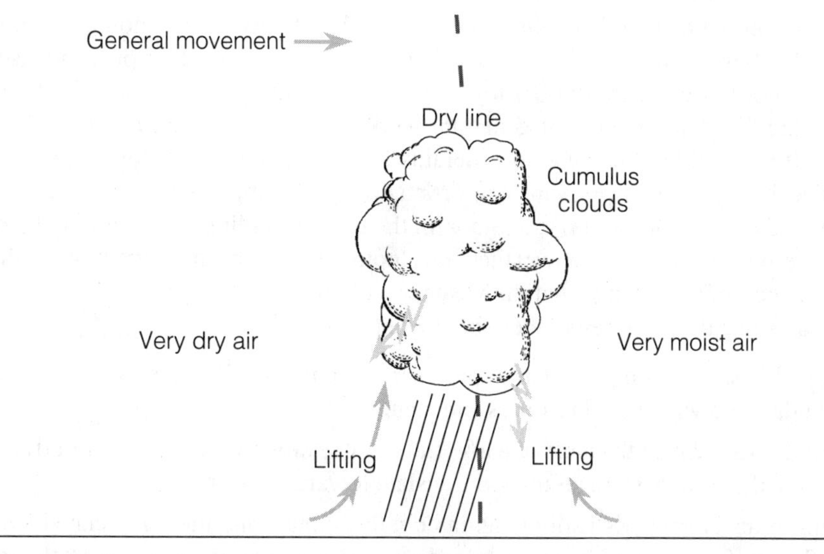

Fig. 4-18. *Storm development along a "dry line."*

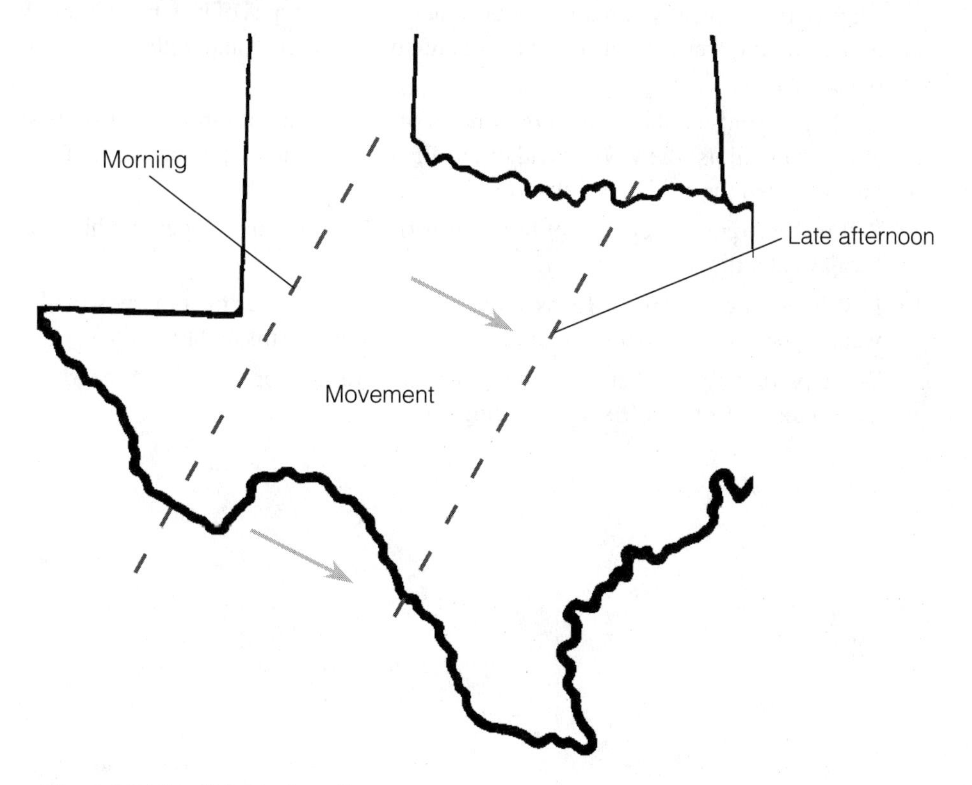

Fig. 4-19. *The "Marfa front."*

Static air pressure alone is not an indicator of what the weather is going to do. A reading of 30.10 might be high air pressure, but if it's lower than the air pressure elsewhere, then moisture is going to flow my way. Instead, what is important is the trend in air pressure. Here an axiom will generally work: dropping pressure leads to worsening conditions, while rising pressure heralds better weather. Even this isn't foolproof, however, because air pressure will drop slightly as temperature rises on a nice afternoon, and air pressure will rise again with the night's cooling. Air pressure, then, is a measure of the potential for weather development, but you must look at all the other factors before predicting what the weather will do.

Here are some things to remember about fronts:

- Frontal systems form off of low-pressure systems. You'll never see a frontal boundary drawn from a high-pressure center.
- Cold fronts form on the west of a low and revolve around the south; warm fronts form in the southeast quadrant and translate toward the east.
- Cold fronts characteristically contain cumulus clouds and their associated hazards; cold fronts are also unstable and generate storms; the hazards of slower cold fronts might be obscured in stratus clouds.
- Warm fronts typically contain stratus clouds and their types of hazards, although a fast-moving warm front might also contain mixed icing and embedded thunderstorms.
- Occluded fronts contain the worst aspects of warm and cold fronts and are best avoided by pilots; the worst weather usually occurs at the time and point of impact between the cold and warm fronts.
- Stationary fronts can create cold- or warm-front hazards over a geographic area for days at a time.
- Dry lines often form in Texas and the Great Plains during otherwise calm weather periods; dry lines can create strong thunderstorms and tornadoes.
- When predicting weather, the state of air pressure (low or high) is not nearly as important as the trend (rising or falling).

5
Wind

WE'VE ALREADY DISCUSSED HOW WINDS DEVELOP AS A CONSEQUENCE OF unequal heating and the resulting differences in air pressure. Air flows clockwise away from high pressure areas and spirals counterclockwise into regions of low air pressure, bent by the Coriolis effect of the earth's rotation. We've also covered the general circulation of the atmosphere: rising air over the equator, an area of descending air from the south, rising air to the north in the mid-latitudes, and descending air in the cold polar regions. Now we'll look at some more general rules of thumb for the wind.

SURFACE WINDS VERSUS WINDS ALOFT

The general rules of wind movement change close to the ground. Surface friction tends to cancel out the effect of the Coriolis force, bending the winds 20 to 40 degrees towards the center of the low. This will always be a directional change towards the wind's own left. The greater the wind speed, the more inertia it has, so faster winds bend less than slower ones. The wind speed also slows with surface friction; wind speeds generally double from the ground to about the 1500-foot AGL point (Fig. 5-1).

Why is this important? Close to the ground, crosswind correction figures change. You'll have to modify your crosswind technique continuously on final approach as ap-

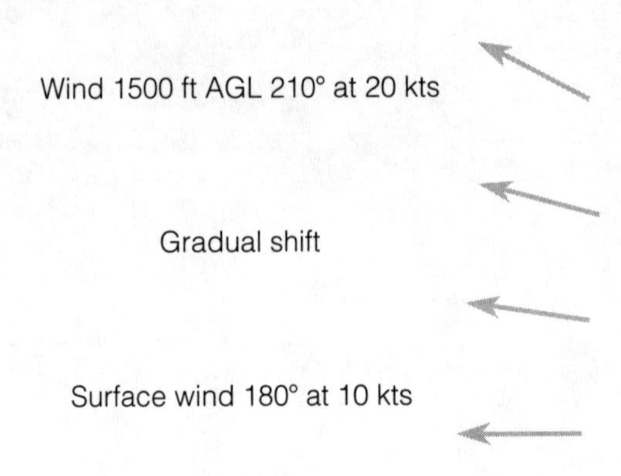

Fig. 5-1. *Wind direction typically shifts 20 to 40 degrees, and wind speed roughly doubles, in the first 1500 feet above ground level.*

parent wind changes. This is especially important in instrument approaches, when you have no outside visual cues about the effect of the wind.

Most instrument approach procedures are flown at about 1500 feet above ground to the final approach fix. The wind correction value that holds you on the final approach course inbound to the fix, then, is not the same wind correction you'll need from the FAF to the field. In practice, regardless of the wind flow at the surface, this change in wind direction will always require you to change heading slightly to the left to maintain precision (Fig. 5-2).

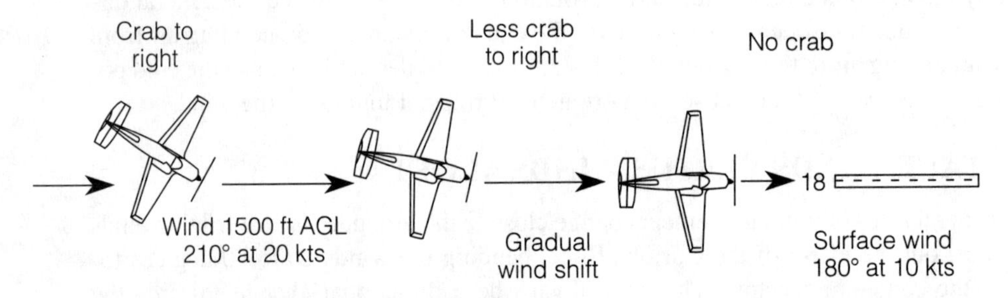

Fig. 5-2. *Changing winds in the lowest level of the atmosphere require a change in wind correction angle on approach.*

Most pilots are competent at following a localizer course from the final approach fix inbound because of the simple presentation of HSI and OBS-style instruments. VOR and RNAV approaches, and those coming soon with loran and GPS, are even easier because the course widths are often wider. Knowing beforehand how the wind will behave, however, makes it easier to anticipate needle movement and maintain courseline right down to minimums.

Where I see the greatest need to understand this wind change in the last part of an approach is when flying NDB approaches. Since fewer and fewer pilots depend on NDB approaches to reach their destinations, there is a mystique about this sort of navigation, and even many instrument instructors spend little time helping their students master the "automatic" direction finder. I know that I was taught, and am guilty for some time of teaching myself, that a pilot should establish a wind correction angle inbound to the final approach fix, and simply continue flying that heading after passing the descent point. That's why, I think, many people feel the NDB can't be flown with precision: simply holding a constant heading will get you in the general vicinity of the airport, good enough if the ceiling and visibility are relatively good, but it won't suffice in an approach to minimums.

Let's say you're flying an instrument approach, and you determine you need a heading slightly right of the inbound course to compensate for the wind. Expect to cut that heading correction roughly in half as you descend to the ground. If there's no apparent need for crosswind correction, anticipate the requirement for a slight left correction from the fix inbound. Carrying a left wind correction towards the marker? As you near the airport, you'll need to roughly double that figure, as a rough estimation for changing winds (Fig. 5-3).

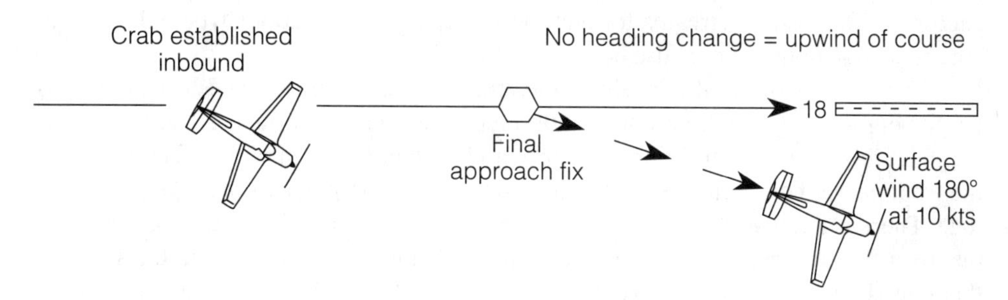

Fig. 5-3. *The effect of maintaining a constant heading on descent through the lowest layer of the atmosphere.*

The faster the wind speed, as we've already seen, the less directional change there will be in the lowest levels of the atmosphere. Therefore, a gust of wind, which lessens the directional shift, will require you to turn to the right (lessening of a left wind correction angle, or increase of a right), for the duration of the gust. Just expect a change in wind correction values as you near the surface, and anticipate numerous heading changes to maintain precision in gusty conditions.

Higher than 1500 feet above ground level, the wind turns and accelerates towards the average southwest-to-northeast, 60-knot wind characteristic in the mid-latitudes at the tropopause height. The wind direction changes fairly consistently with altitude, bending the shortest way around the compass to the SW-NE flow. For instance, if the wind near the ground is out of the southeast, it will be southerly, and then southwesterly, with an increase in altitude. If the surface flow is out of the northwest, winds aloft will be westerly and finally southwesterly with a climb towards the tropopause. Speeds will increase steadily in most cases, although easterly winds might still exist as high as 15,000 feet above ground in strong pressure systems (Fig. 5-4).

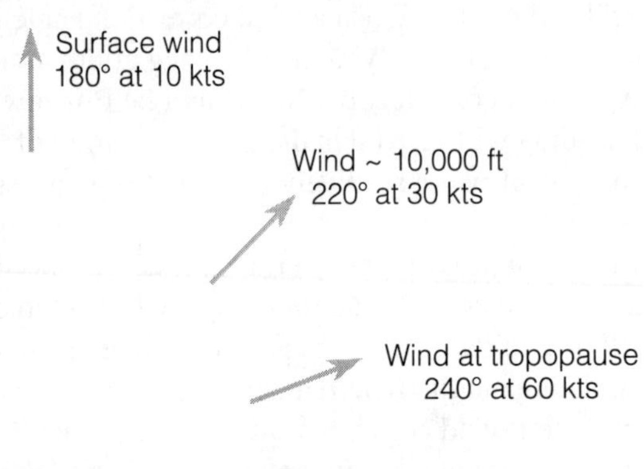

Fig. 5-4. *Wind speeds gradually increase with altitude; wind direction progresses the shortest way around to the prevailing southwest-northeast flow.*

THE JET STREAM

The jet stream is a channel of high-speed wind at altitude, by definition travelling in excess of 60 knots. Jet streams form at the tropopause, in locations where there is an abrupt change in the tropopause height.

When we looked at the levels of the atmosphere earlier, we said that the tropopause height was generally around 60,000 feet over the equator and 20,000 feet over the pole, this polar height dropping to around half that figure in winter. The height of the tropopause, however, does not gently slope from the polar to the equatorial altitude. There is a rather abrupt east-west boundary between the Caribbean air mass and that over the continental United States, marked by the Gulf Coast. Air to the south of this line is generally hot and expands, while air to the north is cooler and contracts. This generates a sudden change in the tropopause height, a break, roughly at the southern border of the United States. Similarly, in the northern plains states near the Canadian border, air temperatures generally cool rapidly with latitude, creating a second break. Jet streams form in the venturi effect of these breaks; we have a semipermanent "equatorial jet" near the southern U.S. border and a "polar jet" in the north (Fig. 5-5).

The jet stream indicates the general flow of weather systems. It is also the prime source of moisture available for condensation and eventual aviation weather hazards. Think of the jet stream as the way moisture is ducted over the continent, and high- and low-pressure systems as the means by which that moisture is distributed.

If you look at popular weather outlets, as you'd find in your local newspaper or on cable television, you'd be led to believe that jet streams are continuous rivers of air, meandering over the continent. In reality, jet "cores," as they are known, tend to be narrow and short, usually a few miles wide and 300 to 1000 miles long. The riverlike flow on TV shows you where the moisture is coming from and where surface systems are likely to blow, to be sure; however, the weather hazards directly associated with the

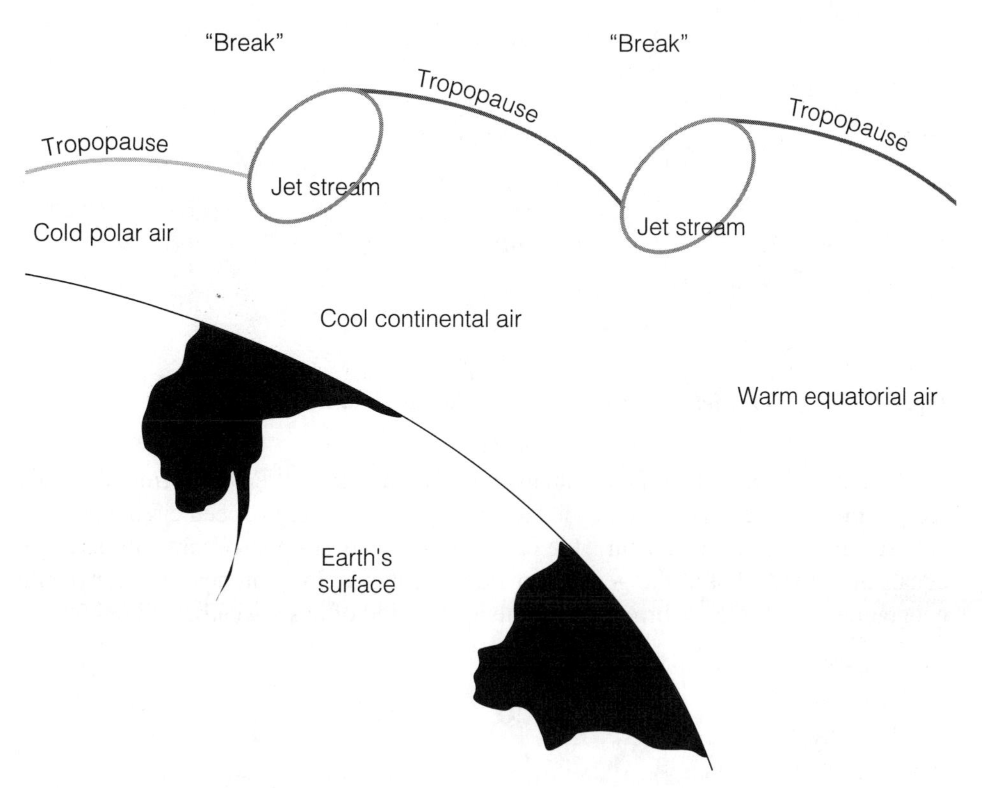

Fig. 5-5. *The jet stream forms at "breaks" in the tropopause.*

jet stream (turbulence and, to a lesser extent, thunderstorms) are limited to the locations of the developed, high-speed cores. Later we'll discover how to identify these locations and anticipate the severity of the associated hazards.

As the jet flow snakes across the continent, it sometimes forms "dips" to the south or north. These bends in the jet stream usually mark the core location; the tighter the turn, typically, the stronger the wind flow aloft. They also provide a clue to the three-dimensional weather picture required to understand the effect of pressure systems. A dip southward in the jet indicates a "trough" (usually spelled "trof" on the charts), an area of low pressure trapped in the upper levels of the troposphere. We've seen how troughs (sometimes called an "upper-level low" or an "upper-level disturbance") increase instability and exaggerate any lifting action below to increase the severity of low-level systems. Thunderstorms, precipitation, and cloudiness might result beneath a trough when surface maps plot only high pressure. Conversely, a northward bend in the jet stream shows a "ridge," an area of high pressure aloft; cold, descending air from these systems increases stability and can negate the effects of low pressure systems at the surface (Fig. 5-6).

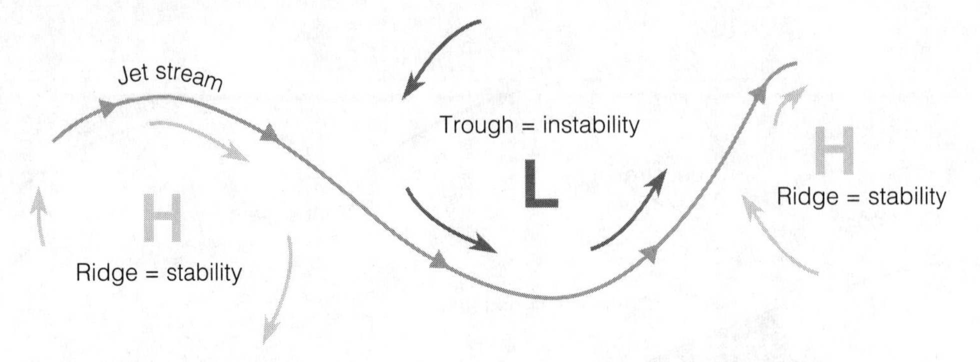

Fig. 5-6. *Bends in the jet stream flow provide clues to weather near the surface.*

Sometimes a trough aloft can intensify to the point that it separates from the main jet stream flow, creating what's called a *cutaway low*. The high-speed circulation of a cutaway low draws in moisture for condensation and storms; because its strength equals or exceeds that of the general air flow, a cutaway low can nearly hover over a geographic area for days, bringing an extended period of nasty weather (Fig. 5-7).

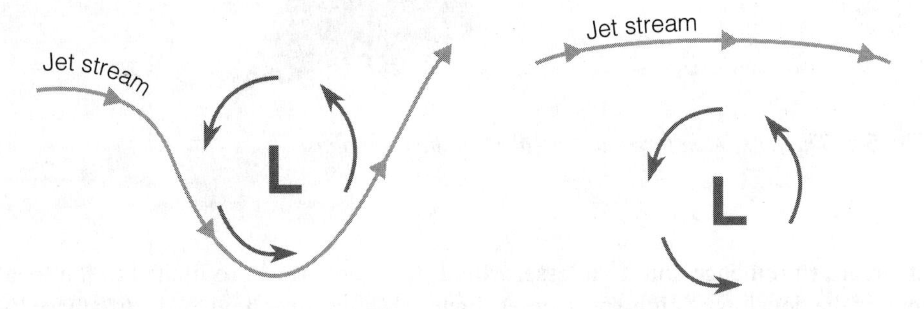

Fig. 5-7. *Formation of a "outaway low." A strong low-pressure area aloft generates jet-speed winds (left), breaking away from the main jet flow (right).*

Although the jet stream often heralds adverse weather, it isn't always marked with clouds or precipitation. It might be invisible while creating moderate or even severe turbulence. The area of maximum turbulence typically forms beneath a jet core and extends as much as 20,000 feet below the tropopause altitude, spreading 50 miles or more on the north, cold side of the jet (Fig. 5-8). This jet-core instability is what's usually called *clear air turbulence*; in winter, when the tropopause height in the continental United States can drop to as low as 10,000 feet, this potentially severe turbulence can affect even the lowest-performing airplane. In chapter 9, we'll cover more about jet streams, including how you can plot the location of developed jet stream cores to anticipate their effect on the safety of your flight.

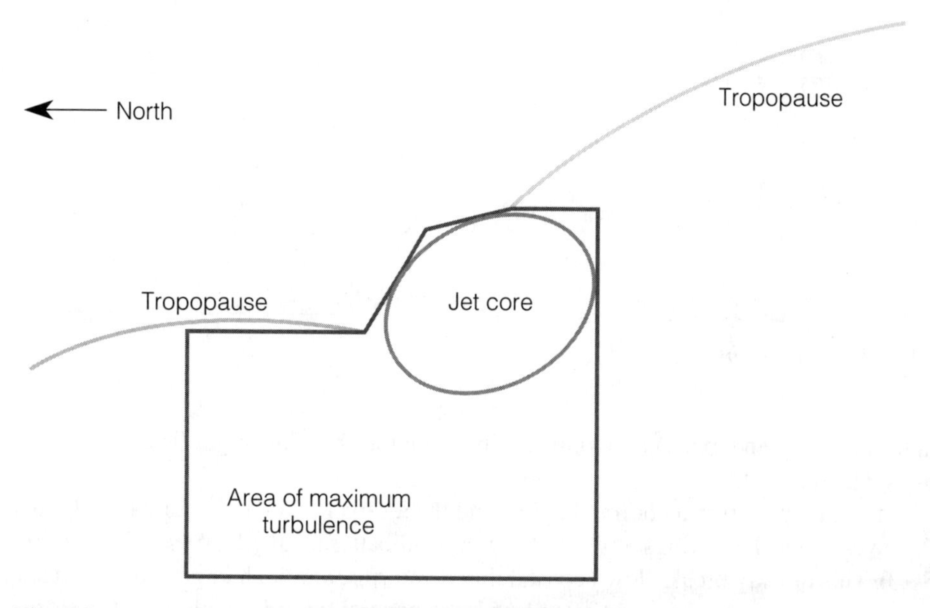

Fig. 5-8. *Area of maximum turbulence associated with a jet core.*

LOCAL WINDS

So far we've looked at the general flow of the wind on a continental and even global scale. Local geography and circumstances, however, can alter this wind flow significantly on a local level; you can anticipate wind flows away from weather-reporting points by knowing some basic rules for local wind flow.

Land and sea breezes

The presence of bodies of water, even as small as lakes, can dramatically and predictably alter wind patterns. As we've seen, water tends to moderate temperature swings. Let's look at a typical coastal area to illustrate the effect of bodies of water on the wind.

Exposed to sunlight, the air temperature over land might vary as much as 20 to 40 degrees Fahrenheit during the day in coastal regions, while the moister air over water changes 5 degrees or less. As the land heats during the day, then, air pressure drops relative to that over the water. Air flows from high to low pressure, creating what's called the *sea breeze* (Fig. 5-9). If the air over land is reasonably hot, it's usually unstable, and this influx of moisture during the day creates the typical cumulus cloud cover and eventual thunderstorms that are common to places like Florida, even on an otherwise beautiful afternoon. If the land surface remains cool, only a few degrees above that over water, the air tends to remain stable, and a thick fog will form in the mid- to late afternoon in locales such as San Francisco. Flying into a coastal airport? Try to arrive

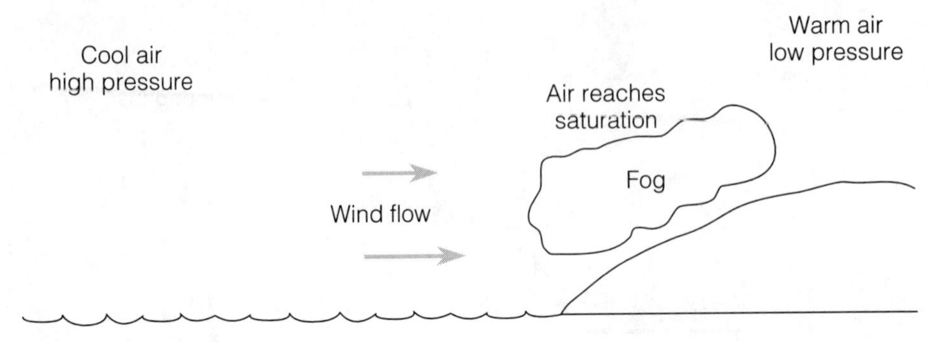

Fig. 5-9. *The sea breeze.*

in the morning, and expect a mounting wind from the shoreline regardless of what stations inland report.

At night, radiational cooling begins, and the air over land cools more rapidly than that over water. This brings a reversal of the wind pattern, called the *land breeze* (Fig. 5-10). This quickly breaks down coastal thunderstorms, although large cells might survive to blow from a west coast inland with the general wind flow aloft, or the storms might travel out to sea to affect areas east of a land mass. The land breeze also pushes fog banks seaward, although a very thick fog might cut off heating over land, negating the land/sea breeze pattern and limiting visibility for days. Is your seaside destination fogged in this afternoon? Chances are it might clear out overnight.

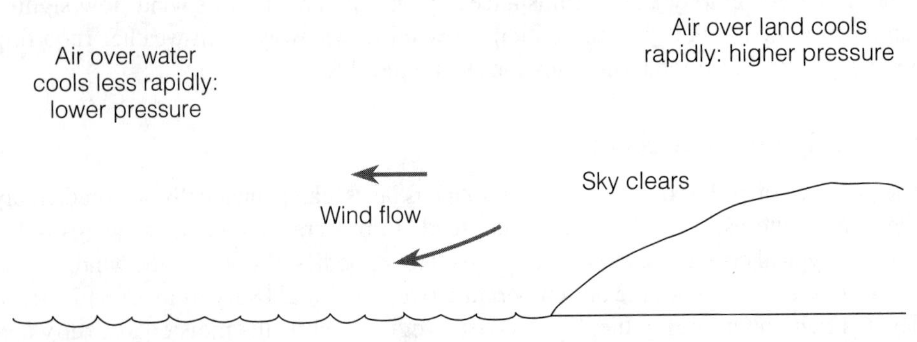

Fig. 5-10. *The land breeze.*

Valley and mountain breezes

Mountainous areas develop local wind flows similar to land and sea breezes. In the daytime, as heating begins, lower-altitude valley air heats and rises. This creates the valley wind, a flow upslope despite the general air circulation. Moisture in this rising air might condense along the ridges, forming the late-morning *cap cloud* familiar to observers in mountainous regions. This cap cloud might well build into a thunderstorm and detach from the peaks with the high-altitude wind flow generally east in late afternoon (Fig. 5-11).

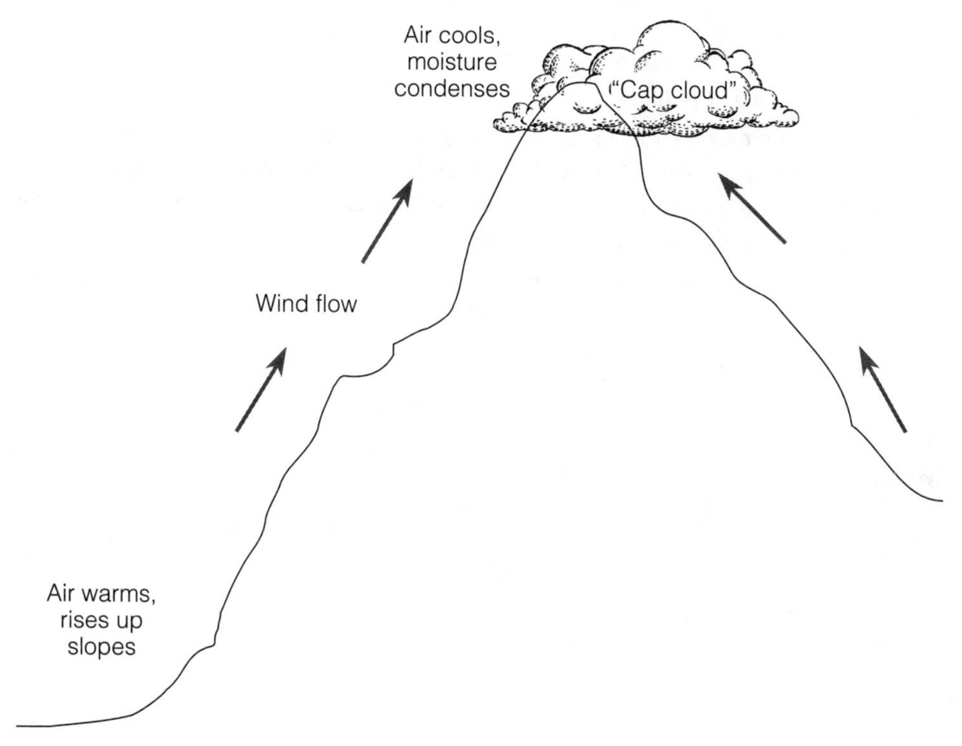

Fig. 5-11. *The valley breeze.*

As the sun goes down, slopes cool rapidly, and cold air flows downhill to become the *mountain breeze*. This explains the rapid clearing of ridge lines just before sunset, often seen in Colorado and southern California. As this air descends, the valleys themselves might act as venturis, accelerating the wind to speeds above 80 miles an hour; the air heats as it descends.

If conditions are right, a hot, dry wind gusts into valleys and the plains beyond; such wind is called the Santa Ana wind in southern California and the Chinook along the Front Range of the Rockies (Fig. 5-12). When flying in mountainous areas, then, anticipate the mountain and valley winds when choosing a flight path or a runway to use; wind speeds and directions can vary dramatically, even over short distances.

Here's a review of some tenets of wind flow:

- Winds typically turn 20 to 40 degrees in direction, and they roughly double in velocity in the first 1500 feet of the atmosphere above ground level.
- From 1500 AGL up, winds normally change speed and direction to the average 60-knot, southwest-to-northwest air flow common at the tropopause.

- Jet stream winds provide moisture and act as steering currents for low-level pressure patterns.
- Sharp dips in jet streams to the south indicate the presence of a trough aloft, which increases instability and tends to increase the severity of surface systems; sharp curves to the north point to the location of a ridge, which increases stability and typically reduces the severity of weather patterns beneath.
- The aviation weather hazards associated with the jet stream, most notably clear air turbulence, form not under the general jet stream flow but instead beneath and north of jet stream cores. Cores are areas where the wind flow at the tropopause exceeds the typical 60-knot pattern.
- Near mountains or along a shoreline, local winds can differ dramatically, yet predictably, from those at nearby reporting points .

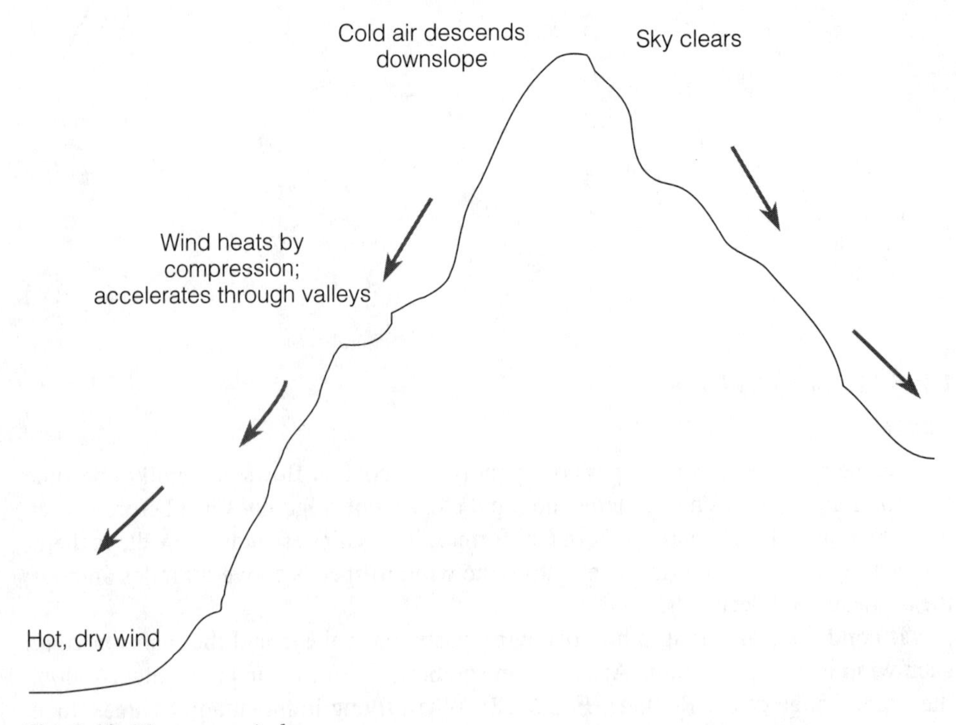

Fig. 5-12. *The mountain breeze.*

6
High-altitude weather

WE'VE ALREADY COVERED QUITE A BIT ABOUT PECULIARITIES OF THE UPPER atmosphere. Now let's review those details for pilots of airplanes that are capable of reaching the tropopause and beyond. The tropopause is the boundary between the lower troposphere and the stratosphere. The troposphere is where most atmospheric heating, condensation, and weather hazards take place, and the stratosphere is the cold, dry layer of air above the troposphere. As with any other boundary between relatively warm, moist air masses and drier, colder ones, the tropopause is a region of turbulence and wind flow.

Jet streams form at breaks in the tropopause, which average around 35,000 feet over the continental United States in summer and as low as 17,000 feet in winter; the biggest hazard associated with the jet stream is clear-air turbulence. This turbulence is greatest at and just below the tropopause in developed jet cores, and the turbulence extends 50 to 100 miles on the north, cold side of the jet. Cirrus clouds might mark the southern edge of the jet stream if enough moisture is present.

Penetrating these or any other high-altitude, cirrus-type clouds usually represents little hazard. Because these clouds are composed mainly of ice crystals, there is little chance of airframe icing, and any that does form will likely be a trace accumulation of rime ice. The biggest problem you're likely to experience flying through these clouds is precipitation static, that electrical charge built up on your airframe as it brushes against ice crystals. Check your airplane's static wicks before flying at this height; an improperly grounded airplane might experience navigation and communications outages due to precipitation static. Lorans and, to a lesser extent, transponders seems to be most susceptible to this problem.

Thunderstorms can easily top the tropopause height; if you see buildups above the altitude, call that storm severe. Watch out for icing and turbulence in clouds, and there's a good chance of hailstones anywhere within 20 miles of the cloud and within a few thousand feet above its top.

7
Seasonal weather variations

WE'VE BRUSHED UP ON THE EFFECT THAT DIFFERENT SEASONS HAVE ON weather patterns. Now let's review the general rules.

SUMMER

Low-pressure systems prevail in the continental United States in summer. In other words, if no other systems blow in from elsewhere, the air over land tends to get hot, creating low pressure systems as the "natural" state. Warm to hot weather, however, extends well north even into Canada, meaning that there's usually little temperature contrast between areas of low pressure and those that are relatively higher. Wind speeds, therefore, are reduced, lessening the severity of most storms and other hazards. This is less true further north in the United States, closer to areas of greater hot/cold contrast. In fact, the "tornado belt" shifts north from the Great Plains to the Great Lakes region in late summer.

Typically, atmospheric pressure stays within about a tenth of an inch of standard in summer, limiting wind speeds and the passage of fronts over a geographic point. On average, expect a single frontal passage per week at any given location. Hot air can hold a lot of moisture, so condensation is limited; there's little IMC weather, and the precipitation that does form is usually due to lifting action and its associated cooling, creating isolated thunderstorms.

In summer the polar jet stream has retreated far to the north, leaving the continental states under the influence of a single, equatorial jet. The tropopause is high enough in summer that jet stream turbulence hazards aren't a factor for naturally aspirated or low-altitude airplanes; watch for storm development in the lifting action under the jet.

WINTER

In contrast, the "natural" pressure state in winter is the high. Unless warm, southern air blows in, exposed areas cool with the lengthened night. There is a greater contrast in pressures in winter when compared to summer, however, so wind speeds tend to be higher. Add to that the lowering of the jet stream and the influence of the polar as well as equatorial jet, and you have stronger winds and two or even three frontal passages per week for a given spot on the map. Mountain waves are more common in winter because of the lowered jet stream as well.

Cold air can't contain much moisture, so widespread areas of fog and "low-IFR" weather are common when winds flow from a moisture source. In the dead of winter, however, weather is usually benign simply because the extremely cold air transports very little moisture. Of course, this is most true in northern, inland parts of the United States.

SPRING AND AUTUMN

The change of seasons brings the most violent weather. It's no coincidence that most big blizzards and tornadoes occur in spring, and hurricane season is in the fall. These are the times of year when the contrasts are the greatest between warm, southern air and cold, northern air, and these two contrasting air masses collide in the United States. Pressure patterns move rapidly, bringing severe weather, then stall out in warmer climes (where the contrasts are lessened) to bring drenching days of embedded storms. Expect a lot of heavy weather as the seasons change, gradually transitioning to the "natural" winter or summer states as time passes; factor in latitude to anticipate the climatic patterns for a given region.

The next part of this book covers the four categories of aviation weather hazards: thunderstorms, turbulence, reduced visibility, and ice. For each of these hazards, we'll look at what causes them to form, how to anticipate where they'll be found, and how to devise a plan of action to avoid the hazard. We'll also cover aviation weather products. Some are well-known and routinely provided in aviation weather briefings; some are more esoteric, yet very revealing in predicting the intensity and likelihood of hazards. Then we'll discuss methods of avoiding unfavorable weather, or at least limiting your exposure to risk; finally, we'll review techniques for getting yourself out of trouble should your plan for avoidance fail.

Throughout the remainder of this book, I'll reference aviation weather products, some taken from EA-AC-0045B, the government's Aviation Weather Services advisory circular; others are actual "hard copy" from Flight Service Station printouts. I have not attempted to teach complete decoding of these products; for more assistance in decoding, I suggest Terry T. Lankford's *Pilot's Guide to Weather Reports, Forecasts and Flight Planning*, also by TAB Books, and Advisory Circular 0045B.

PART 2

Aviation weather hazards

8
Thunderstorms

ONE OF THE MOST SPECTACULAR WEATHER PHENOMENA, THUNDERSTORMS develop rapidly, move quickly across the ground, and can contain all three of the other aviation weather hazards. Turbulence, reduced visibility near the surface, and airframe ice all can and do exist in the body of a thunderstorm. We'll now cover the life cycle of a thunderstorm, the exact nature of the hazards thunderstorms present to pilots, weather products that warn of a storm's development, and techniques for avoiding thunderstorms and surviving an encounter should attempts at avoidance fail.

There are three elements required for a thunderstorm to develop:

- A source of moisture.
- Some sort of lifting action.
- Unstable air.

From a pilot's perspective, there are two types of thunderstorms: frontal storms and air-mass storms. *Frontal thunderstorms* are those, as the name suggests, that occur along the boundary of dissimilar air masses (Fig. 8-1). *Air-mass thunderstorms* form when some nonfrontal lifting action exists, such as convection in the heat of afternoon, orographic lifting (where sloping terrain forces air upward), passage of the

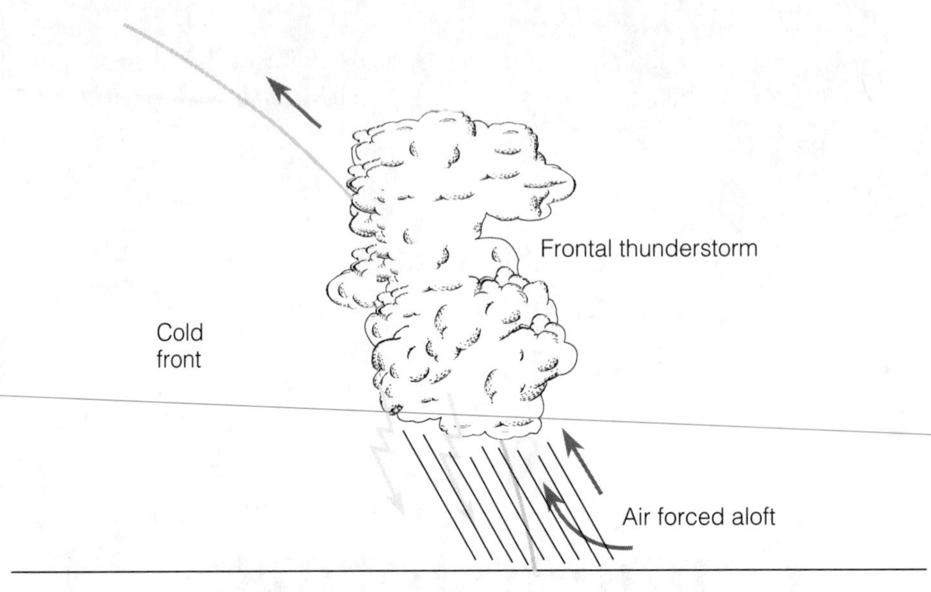

Fig. 8-1. *A frontal thunderstorm.*

jet stream (which acts like a giant vacuum cleaner, sucking air upward), or along coastal areas where the sea breeze pushes air inland to heat and rise over the land (Fig. 8-2). The main difference between frontal and air-mass thunderstorms is that frontal thunderstorms can be self-renewing, lasting several hours and becoming quite violent; air mass thunderstorms are no less severe, but they typically pop up and dissipate in a matter of 20 minutes to half an hour.

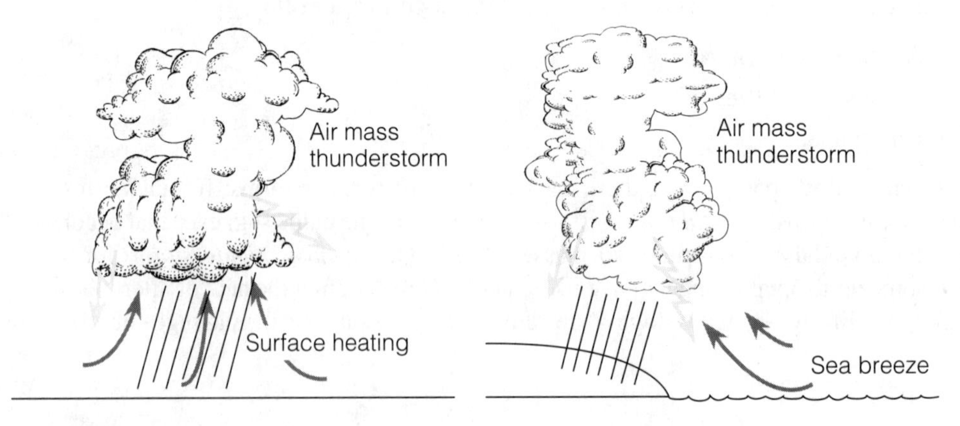

Fig. 8-2. *Air mass thunderstorms created by convection (left), or local wind patterns (right).*

Figure 8-3 depicts the beginnings of a thunderstorm. With all the elements in place (moisture, lifting action, and instability), the storm begins in the cumulus, or updraft, stage. At this point, vertical updrafts can exceed 5000 feet per minute, beyond the ability of almost all aircraft to overcome. The primary danger to pilots in the updraft stage is the wind shear that might exist on the boundaries of the rising air columns.

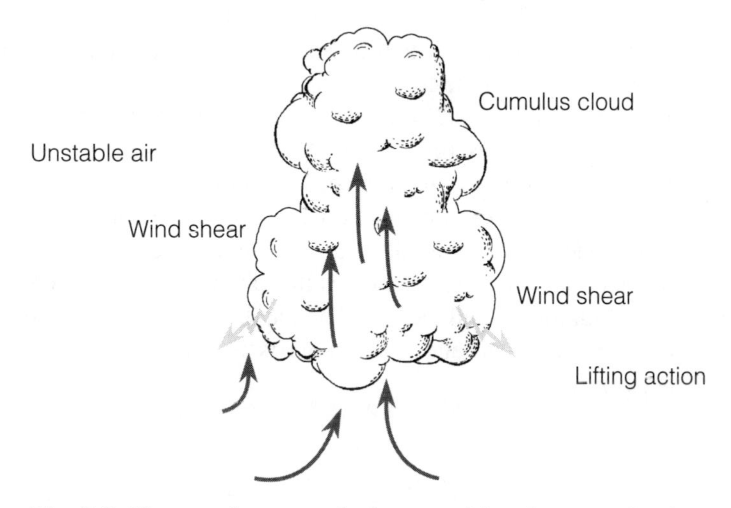

Fig. 8-3. *The cumulus, or updraft stage of thunderstorm development.*

I was flying a Cessna 120 over central Missouri several years ago. It was a bright, sunny early afternoon, a little bumpy down low, and a single, small cumulus cloud lay between me and my home field. I underflew its base by several thousand feet when suddenly I was being sucked upward; at idle power and while pushing full forward on the yoke, the vertical speed indicator of my 85-horsepower trainer was pegged on the 4000-foot-per-minute mark. Just before I was about to penetrate the base of the still innocent-looking cloud, I flew from beneath it, only to find myself screaming at the earth in a 4000-foot-per-minute-plus dive. Shortly afterward I was on the ground at my home airport, looking at a monster thunderhead a few miles east of the field.

Interestingly, at this early yet dangerous phase of a thunderstorm's life, it usually isn't visible on radar. Radar doesn't detect storms or turbulence directly; instead, radar reflects off precipitation. There is a strong correlation between heavy precipitation, thunderstorms, and turbulence, but that doesn't help in the early stages of a storm's development. The cumulus-stage thunderstorm will, however, show up on an electrical discharge detector like a Stormscope or a Strikefinder because the friction between rising and still columns of air generates an electrical discharge. Where there is electricity there is turbulence; we'll look at other advantages and disadvantages of both radar and electrical discharge detectors later in this chapter.

The thunderstorm enters the mature stage when precipitation begins to fall (Fig. 8-4). The lifting action that is at the heart of the storm's development can support droplets of water only up to a point. As those droplets collide and combine in the cloud, eventually they become heavy enough that the lifting action can no longer sustain them. They begin to fall. As they fall, raindrops create downdrafts through several methods. First, the droplet displaces air as it falls, creating a slight suction behind the drop, pulling air downward. Second, the drop itself cools the air around it, and cold air descends, adding to the downdraft. If the drop enters drier air it might reevaporate, either in the cloud or beneath it in the form of virga. Evaporation is a cooling process; this further cools the air and increases the magnitude of the downdraft. While the updrafts at this point in the storm's life still might measure more than 5000 feet per minute, the downdrafts themselves might exceed even that figure. When these winds hit the ground, they can spread out into a gust front, a massive wall of wind that creates significant turbulence. A wall cloud, or rotor cloud, signals danger to a pilot. A thunderstorm's primary hazard to pilots, then, is the extreme turbulence on the boundaries of the rising and falling air and the gust front, and the possibility of being pushed to extreme altitudes or into the ground in the updrafts and downdrafts.

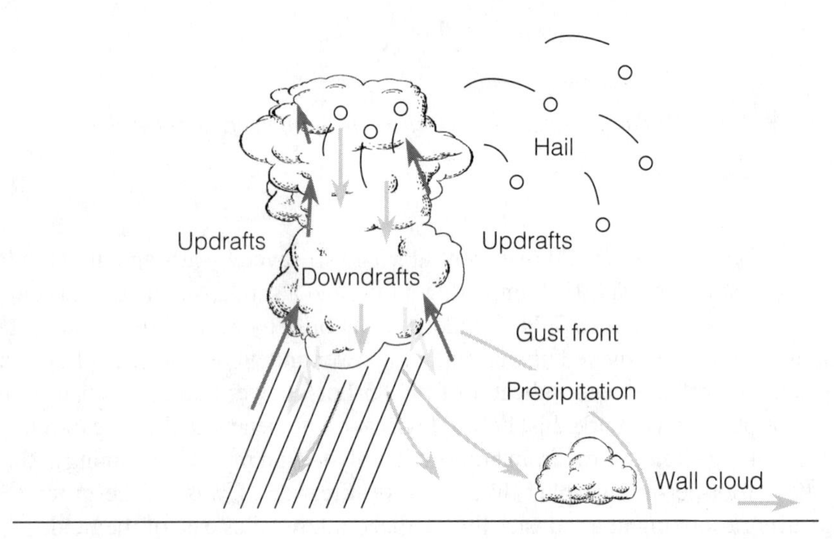

Fig. 8-4. *The mature stage of thunderstorm development.*

Of course, lightning strikes also concern the pilot, but a well-grounded airframe, protected with adequate static wicks for electrical discharge, should experience little difficulty following a lightning strike. I do know, however, of a King Air crew that lost about 2 feet of wingtip and a host of avionics to a lightning strike after some aftermarket wingtip modifications. If you have your airplane painted, maintained or modified, be sure the shop verifies the integrity of your airplane's grounding before flight.

Recent research indicates that lightning tends to concentrate in two levels of a thunderstorm: within about 2000 feet above and below the freezing level and in the zone from 36,000 to 46,000 feet. Heavy rains might mar visibility, and there is also research underway investigating the hazards of airflow disruption brought on by heavy rain.

If temperatures are right, rising droplets of water might freeze into hail; in severe storms, the process might repeat until hailstones become large enough to heavily damage airplanes. Hail has been known to be flung as much as 20 miles from the boundary of a storm cloud, especially on the downwind side. Another hazard, large droplets of water suspended in a massive cumulus formation, warn of heavy amounts of clear ice.

Tornadoes, vortices of extremely low air pressure, present a serious turbulence and structural hazard. Recent evidence suggests a phenomenon known as a *horizontal tornado*, an invisible exchange of tornadic winds between storm clouds spaced less than 30 miles apart. The recent as yet unexplained crash of a jet airliner near Colorado Springs might have been the result of a horizontal tornado encounter.

All of these downdrafts quickly take a toll on the thunderstorm cloud and cause it to enter the dissipating, or downdraft, stage (Fig. 8-5). At this point in its life cycle, the cloud is characterized primarily by downdrafts exceeding 6000 feet per minute in intensity. The storm still contains all of the hazards associated with the mature stage, with an increased likelihood of extreme turbulence near the ground. We'll look at the worst-case scenario, a microburst, in the chapter covering turbulence. With little lifting action to sustain the storm at this point, it quickly dies away, but beware of the formation of a new cell if conditions persist.

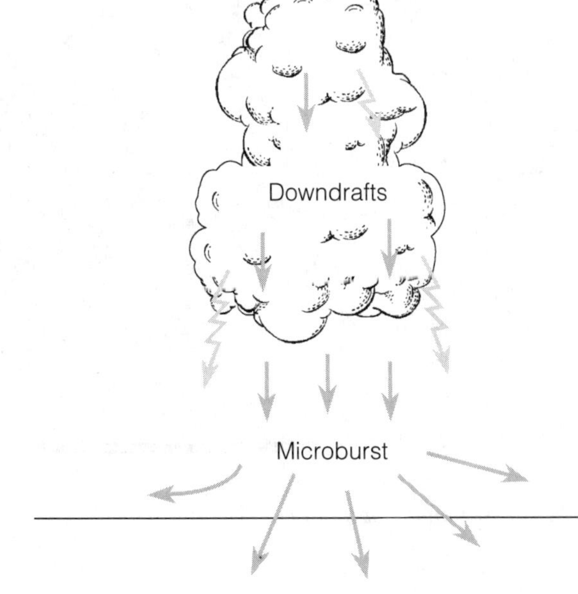

Fig. 8-5. *The dissipating, or downdraft stage of thunderstorm development.*

THUNDERSTORMS ON THE WEATHER CHARTS

There are numerous aviation weather products, or charts, that point to the possibility of thunderstorms. Each has its own advantages and pitfalls; many are well known and routinely referenced in a Flight Service preflight briefing; other, more obscure charts are a wealth of information available only if you ask for them. Let's look at some thunderstorm-related weather products.

Radar summary chart

Figure 8-6 shows the National Weather Service radar network, used to derive the Radar Summary Chart (Fig. 8-7). Notice that the radar network provides almost universal coverage west of the Rocky Mountains, meaning that any precipitation is likely to show up on the chart in those areas. In the mountains and along the West Coast, thunderstorms might go undetected; although most of this region receives very few thunderstorms each year, remember that sections of northeast New Mexico and southeast Colorado experience more storms annually than anywhere outside of the Gulf Coast. You'll need to rely on products other than radar to anticipate thunderstorms in that area.

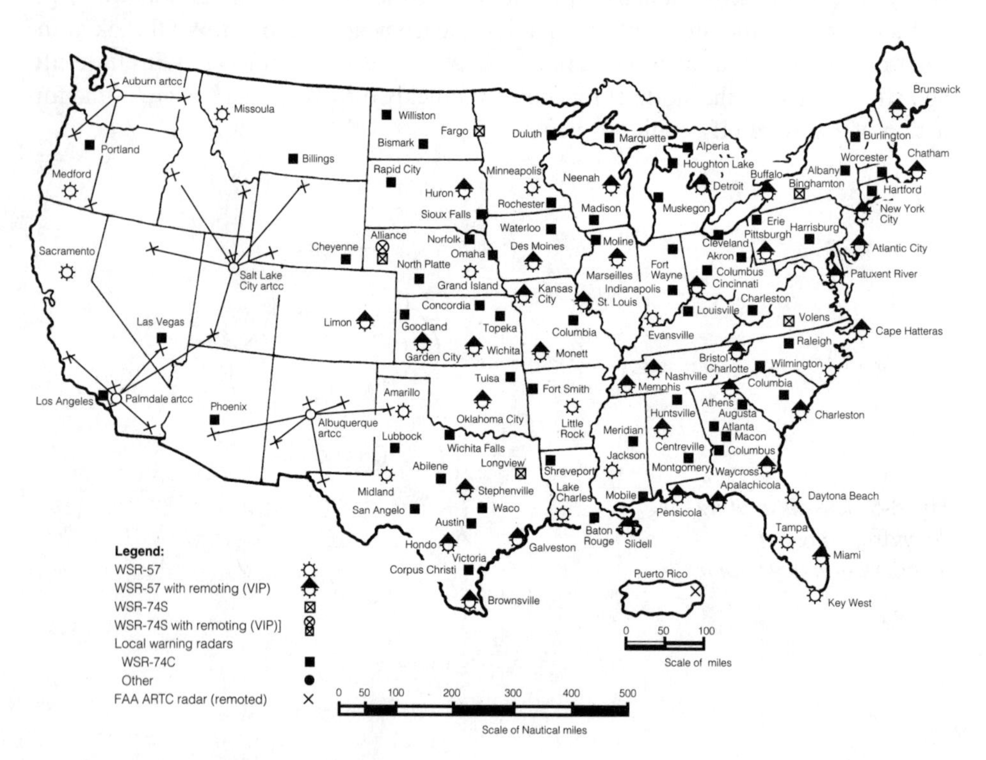

Fig. 8-6. *The National Weather Service radar network (Advisory Circular 0045B).*

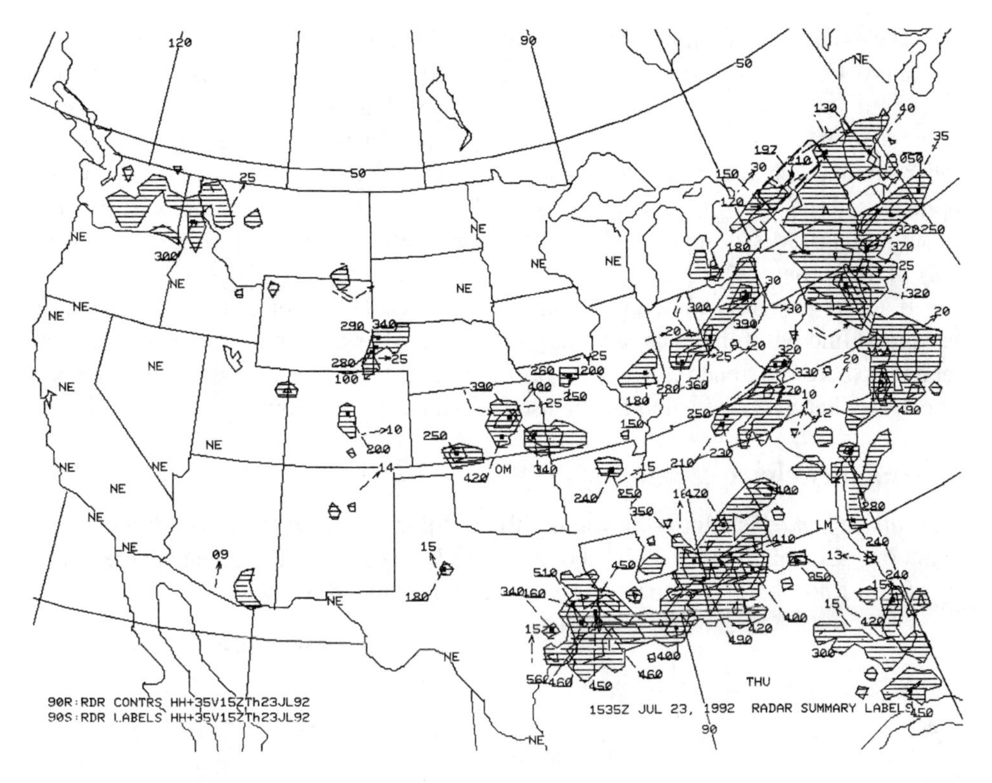

Fig. 8-7. *A radar summary chart (Advisory Circular 0045B).*

The radar summary chart provides reams of information. First, it shows areas of precipitation coverage. Shaded areas indicate level 1 or 2, light precipitation; white areas inside shading show level 3 and 4, moderate precipitation, and dark areas inside the white refer to heavy, level 5 and 6 echoes. Remember that the radar chart does not directly indicate thunderstorm activity. It only indicates precipitation, but there is a strong correlation between heavy precipitation and storm-induced turbulence.

Just because the chart shows all of Ohio filled in with level 1 and 2 shading, for instance, doesn't mean the entire state is receiving precipitation. The radar summary fills in areas between echoes, so it indicates the general boundaries of the rain or snowfall; specific areas within the shading might or might not be receiving precipitation at any given time.

The summary chart indicates the type of precipitation, also, be it S for snow, L for drizzle, ZL for freezing drizzle, R for rain, or T for thunderstorms. A W indicates "showers," as in rain or snow showers; "+" indicates heavy activity while "–" denotes light. The movement of general areas of precipitation is noted, as well as the movement and height of individual cells within the general flow.

The radar summary information is collected by the Severe Storms Forecast Center in Kansas City, compiled into a map, and issued approximately five minutes before each hour. As such, its information might be as much as an hour old when the map is released, and could be two hours old at the end of a map's effective time. Hence, the radar summary chart is meant to be used to show general areas of precipitation development, a trend instrument; use it to say, "Yes, storms are developing, so I'd better avoid northern Mississippi," and not, "There's a cell over Birmingham, so I'll go by way of Montgomery."

If you're using a commercial weather vendor (via DUAT or a facsimile machine), be sure to find out if the radar summary comes from the Severe Storms Forecast Center and is therefore trend information, or if it is "real-time" weather radar from the National Weather Service, which can be used to plan a specific route.

Stability chart

The *stability chart* is a little-known, highly useful indicator of the possibility of thunderstorms. Compiled twice daily, at 00Z and 12Z, it is a measure of the average stability of the atmosphere over specific reporting points (Fig. 8-8).

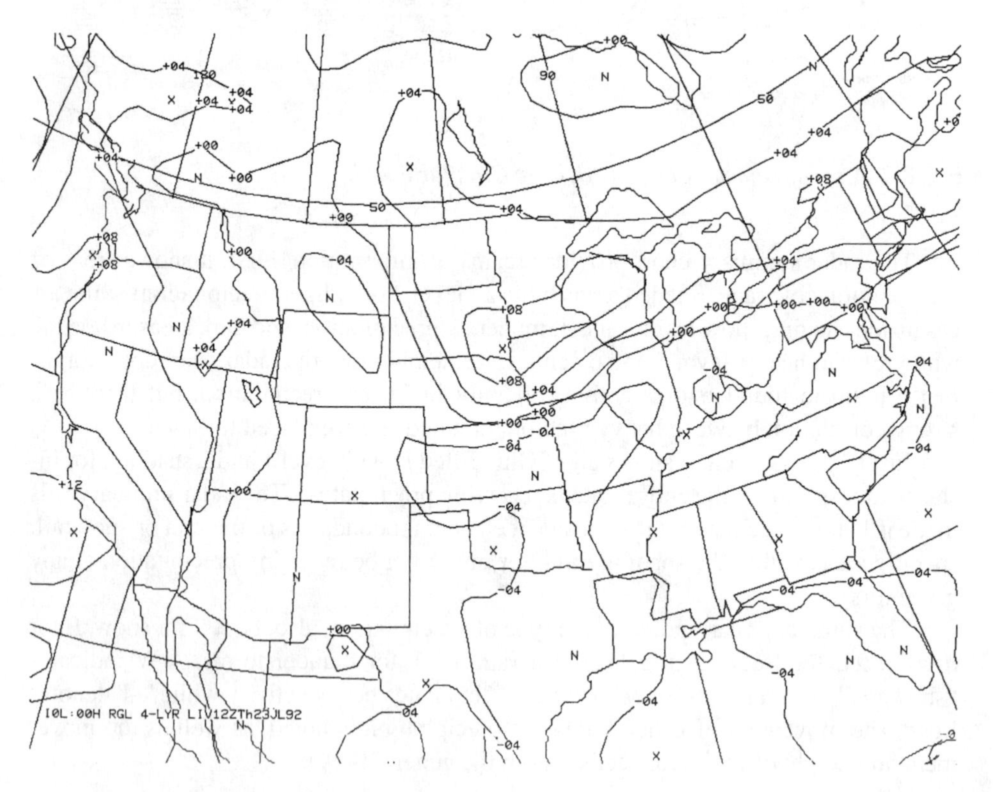

Fig. 8-8. *A stability chart (Advisory Circular 0045B).*

One element of thunderstorm development is instability. If you live in the central United States, you know that from about March to October you'll almost always get a warning about "possible thunderstorms" from the Flight Service folks. If the sky is cloudy or the air is muggy, it makes taking off a nerve-wracking affair unless you have more information. That's when you'll want to ask for the stability chart.

Noted over each reporting point on the chart is what looks like a fraction, one number over another. The number on the top, the "lifted index," is what you want. When the balloons to determine winds aloft are launched, they take the air temperature at preselected altitudes and compare the rate of temperature loss to standard lapse rates. The lifted index is derived as a correction value; for instance, a value of +1 means that, on average, 1 degree would have to be added to existing temperatures aloft for the lapse rate to be standard. In other words, the actual lapse rate is less than standard; the air is stable. The higher the positive number, the more stable the air, so a positive number from the lifted index means that thunderstorm development is unlikely despite the presence of other elements of their formation. Any storms that do form should be small, short-lived, and isolated. Negative numbers indicate unstable air, primed for storm development; the "more negative" the number (–3,–4,–5, etc.), the more severe the ensuing storm is likely to be. Expect severe thunderstorms for values of –4 or greater if moisture is present.

The number on the bottom at each reporting point, the "K index," relates the stability or consistency of the lapse rate. High positive numbers denote unstable lapse rates. High K indices, combined with a negative lifted index, warn of widespread, severe storms.

Planning a flight, you get the stock warning of summertime storms from your DUAT briefing, and conditions outside are partly cloudy. Call Flight Service and ask for the lifted index for your route of flight. If the numbers are positive, you can probably make the trip without encountering thunderstorm hazards. If the numbers are negative, go with a lot of fuel for a diversion, and keep your eyes and ears open for developing storms en route.

Surface aviation reports

Surface aviation reports, the hourly observations made at airports and other reporting points, might also warn of a thunderstorm. If visibility is less than 7 miles, the reason for the reduction in visibility is required to be reported. Look for a "T" for thunderstorms, or "TRW" for thunderstorms and rain showers, in the visibility section of the report. Sometimes you'll find mention of visible thunderstorms, lightning, or towering cumulus clouds in the remarks section of the report as well (Fig. 8-9).

Pilot reports, area forecasts, and terminal forecasts

These three independent reports are illustrated in Fig. 8-10. Pilot reports don't often report thunderstorm activity, but they might; I've even heard a report of tornadoes in a

```
WRB SA 1754 M15 BKN 80 OVC 21/2TRW- 2106/989/RM TWRCU NE
```

Fig. 8-9. *A surface aviation report indicating thunderstorms.*

```
MEM UA/OVR MEM/TM 1750/FL040/TP BE36/IC LT CLR INCL/TB
MDT/LTNG IC/RM ON APCH MEM

CHIC FA 231045
SYNOPSIS AND VFR CLDS/WX
SYNOPSIS VALID UNTIL 240500
CLDS/WX VALID UNTIL 232300...OTLK VALID 232300-240500
TSTMS IMPLY SVR OR GTR TURBC SVR ICG LLWS AND IFR CONDS
AL
NWRN HALF...CIG 10-15 OVC. VSBY 3-5F. SCT RW/TRW. TSTMS PSBLY
SVR.
SRN PTN. AFT 21Z CIG 15-30 BKN/OVC. TOPS 250. CB TOPS 450.
OTLK...MVFR CIG.
SERN HALF...CIG 10-15 OVC. VSBY 3-5F. SCT RW/TRW. TSTMS PSBLY
SVR SRN PTN. TOPS 250. CB TOPS 450. OTLK...MVFR CIG.

MCN FT 231818 8 SCT C35 OVC 2410 OCNL C8 OVC 2R-/TRW
    02Z 6 SCT C20 OVC 5F 2608 OCNL C6 OVC 2 R-F CHC 1 TRW
    08Z C25 OVC 2808 CHC R-
    12Z MVFR CIG.
```

Fig. 8-10. *A pilot report (top), an area forecast (center), and a terminal forecast, all indicating thunderstorms.*

pilot report near Salina, Kansas. Area forecasts will point out the possibility of storm development, as well as expected intensity and movement; the area forecast contains the disclaimer that, "Thunderstorms imply severe or greater turbulence, severe icing, low-level wind shear and IFR conditions," so consider yourself warned.

The Area Forecast is issued three times daily in the continental United States, and the prediction contains an 18-hour synopsis period, a 12-hour forecast period, and a 6-hour categorical (VFR, MVFR, IFR or LIFR) outlook. The terminal forecast is a prediction of the same sort of information listed in a surface aviation report, so if visibilities are expected to be limited by thunderstorm activity, you'll see mention of it in the visibility section of the terminal forecast. These reports are issued three times daily, cover a 24-hour forecast period, and are valid for a 5-nautical-mile area around the reporting point. The word "vicinity" refers to activity within 25 miles beyond that valid area.

Low-level significant weather prognostic

The "prog chart," as it's known, actually consists of four panels on the same page (Fig. 8-11). The two on the left concern the forecast for up to 12 hours from the release time, while the charts on the right are valid from the 12- to 24-hour point. Essentially a pictorial view of combined terminal forecasts, it denotes general areas of expected thunderstorm coverage on the bottom panel for each forecast period. One thing I like about the progs is that they show the expected ground speed of high-pressure systems, which drive cold fronts. Remember, fast cold fronts tend to develop severe storms and squall lines, while slower fronts typically create embedded thunderstorms.

Convective outlook and severe weather outlook

The "c-note," as it's sometimes called, highlights specific areas of expected thunderstorm activity. It's issued at 08Z, 1930Z from February 1 through August 31, and 15Z, and it contains forecasts of slight, moderate, or high risk of thunderstorm development (Fig. 8-12).

Severe weather watch bulletin

This is the same text you'll often see scroll on your television screen when a severe storm watch is issued near your home (Fig. 8-13). Notice how, the more likely severe storms are, the more "plain language" the bulletins become. A watch means conditions are favorable for the development of severe storms over a defined area.

Severe weather watches are issued as needed by the Severe Storms Forecast Center in Kansas City, and as such are disseminated among other places into the Flight Service network. You'll hear about watches, then, in your weather briefing. Warnings, actual sighting of severe storms or tornadoes, however, are issued by local weather services ("The National Weather Service in Garden City has issued a tornado warning," you'll hear on TV). These local reports do not make it into the Flight Service net. A tornado might be bearing down on your destination, and you'll never hear about it from Flight Service, so be very cautious about flying in or near any area where there is a severe weather watch bulletin in effect.

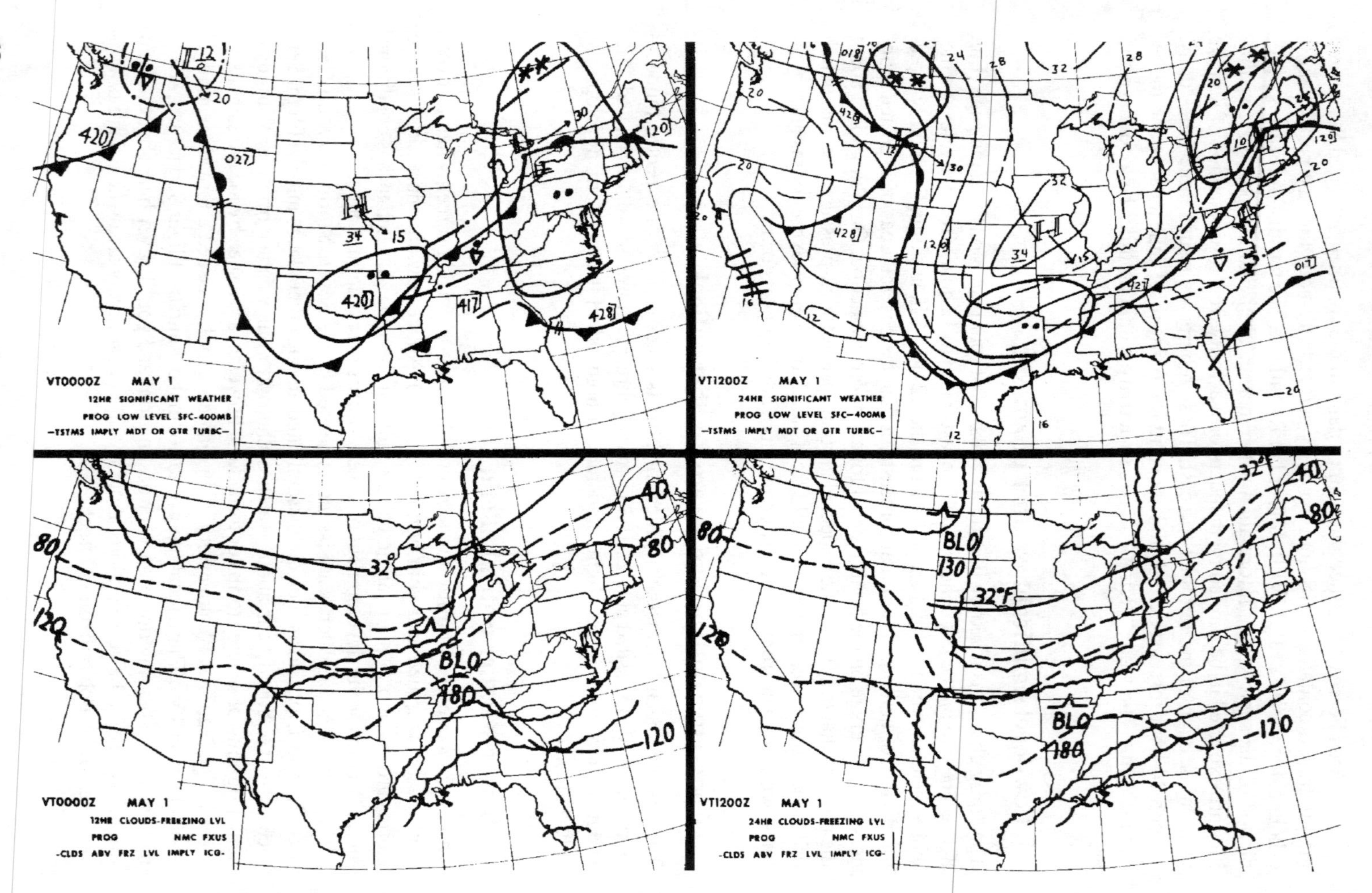

Fig. 8-11. *A low-level significant weather prognostic (Advisory Circular 0045B).*

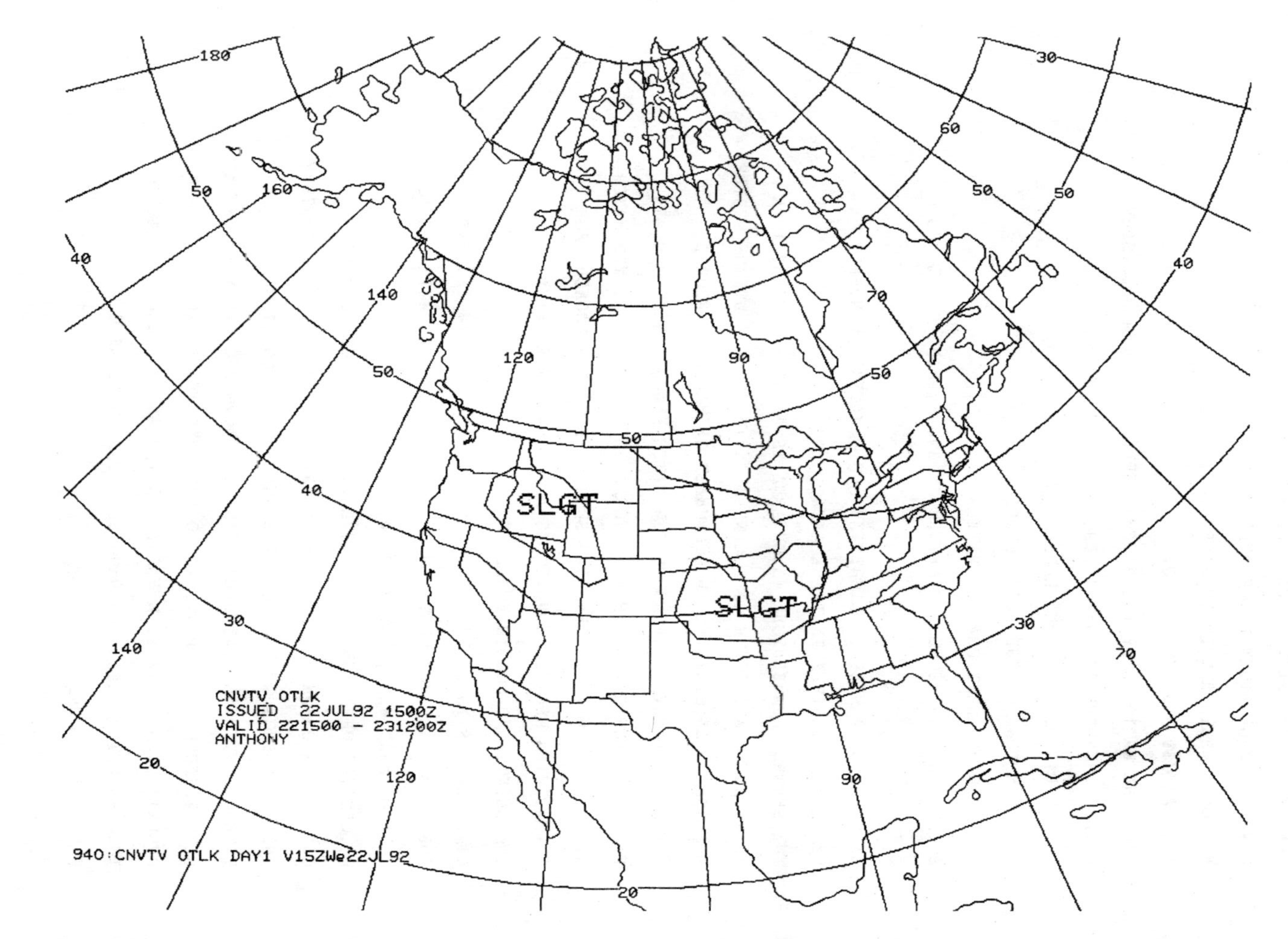

Fig. 8-12. *A convective outlook (top), and a severe weather outlook chart (bottom) (Advisory Circular 0045B).*

```
MKC WW 231755
BULLETIN - IMMEDIATE BROADCAST REQUESTED
SEVERE THUNDERSTORM WATCH NUMBER 17
NATIONAL WEATHER SERVICE KANSAS CITY MO
1155 AM CST WED FEB 23 1994
.A.. THE NATIONAL SEVERE STORMS FORECAST CENTER HAS ISSUED A
SEVERE THUNDERSTORM WATCH FOR
     PARTS OF SOUTHERN ALABAMA
     PORTIONS OF NORTHWEST FLORIDA
     AND ADJACENT COASTAL WATERS
EFFECTIVE THIS WEDNESDAY AFTERNOON AND EVENING UNTIL 600 PM
CST
LARGE HAIL..DANGEROUS LIGHTNING AND DAMAGING THUNDERSTORM
WINDS ARE POSSIBLE IN THESE AREAS.
THE SEVERE THUNDERSTORM WATCH AREA IS ALONG AND 60 STATUTE
MILES NORTH AND SOUTH OF A LINE FROM 15 MILES SOUTH OF MOBILE
ALABAMA TO 25 MILES SOUTH SOUTHEAST OF MARIANNA FLORIDA.
REMEMBER..A SEVERE THUNDERSTORM WATCH MEANS CONDITIONS ARE
FAVORABLE FOR SEVERE THUNDERSTORMS IN AND CLOSE TO THE WATCH
AREA. PERSONS IN THESE AREAS SHOULD BE ON THE LOOKOUT FOR
THREATENING WEATHER CONDITIONS AND LISTEN FOR LATER
STATEMENTS AND POSSIBLE WARNINGS.
C..A FEW SVR TSTMS WITH HAIL SFC AND ALF TO 13/4 IN. EXTRM
TURBC AND SURF WND GUSTS TO 60 KNOTS. A FEW CBS WITH MAX TOPS
TO 530. MEAN WIND VECTOR 22035.
D..CLUSTER OF INTENSE TSTMS OVR THE GULF EXPCD TO MOVE
ONSHORE INTO SRN AL AND NWRN FL NXT SVRL HRS. IR STLT IMAGERY
INDCS PRESENCE OF ENHANCED-V SIGNATURES INDCG HI PTNL OF SVR
TSTMS. CNVTN WILL INTERACT WITH SEWD MOVG SFC BNDRY WHICH MAY
ALSO ENHANCE SVR THREAT. WK LO LVL WNDS AND SHEAR SUG TORNADO
DVLPMT IS NOT LIKELY..HWVR STG INSTBY AND FVRBL UPR DYNAMICS
INDC PSBLTY OF HAIL AND DMGG SFC WIND GUSTS AS CNVTN
PROPOGATES INLND.
...NESS
```

Fig. 8-13. *A severe weather watch bulletin.*

Convective SIGMET

A *convective SIGMET*, or Significant Meteorological Information warning, is a warning of existing or developing convective conditions that might be hazardous to aircraft. They are issued in hourly bulletins around 55 minutes past each hour, and as needed to ensure safety; they are valid for two hours unless superseded. They will be broadcast over navaids, HIWAS, and ATC frequencies when issued, and they will be repeated at 15 and 45 minutes past each hour as long as they are in force. Convective SIGMETs warn of tornadoes, severe thunderstorms, embedded or lines of storms, and thunderstorms of level 4 or greater affecting 40 percent or more of an area of at least 3000 square miles. The convective SIGMET also contains a convective outlook for after the expiration of the two-hour period (Fig. 8-14).

```
MKCE WST 231755
CONVECTIVE SIGMET 14E
VALID UNTIL 1955Z
FL CSTL WTRS
FROM 70WSW CTY-110W PIE
LINE EMBDD TSTMS 20 MI WIDE MOVG FROM 2320. TOPS TO 350.
CDFRT W TN INTO SWRN LA IS MOVG EWD. STG LO LVL JET IS
BRINGING INCRG MSTR ACRS SERN U.S. UPR LVL SPEED MAX OVR LWR
MS VLY WILL HELP SUPPORT INCRSG TSTMS ACTVTY DURG THE PD.
DAYTIME HTG AND OUTFLOW BNDRYS XPCD TO TRIGGER ACTVTY OVT
LAND AREAS.
```

Fig. 8-14. *A convective SIGMET.*

THUNDERSTORM AVOIDANCE

The best thing a pilot can do with regard to a thunderstorm is to avoid it. The best way to avoid a thunderstorm is by visually steering around the cloud. Attempt to stay in clear air, avoiding the cells by at least 20 nautical miles. This puts you out of the reach of horizontal tornadoes and hail thrown out of the storm's side (if you avoid all cells by at least 20 miles, you won't be between cells closer than 30 miles apart). Don't fly under the anvil of a storm cloud, again to avoid hail; hail doesn't form in the anvil, but an anvil is indicative of strong horizontal winds that might propel a hailstone.

Avoid flying under a storm cloud. For a long time the conventional wisdom was that, if thunderstorms threatened, the safest course of action was a VFR flight beneath the bases. This has been proven to be incorrect, with increased knowledge of the dynamics of downdrafts, gust fronts, and microbursts. Avoid flying under the storm.

There are three types of technology designed to track storms: ground-based radar, airborne radar, and electrical discharge detectors. Ground-based Air Traffic Control radar is not tuned for weather reception. New-technology ASR-9 radar is designed to plot weather as well as airplanes, but, as of this writing, technical problems are delaying its introduction. Approach control radar frequencies are a little better at weather detection than those employed at Center; your best bet, however, is to leave controller frequency and talk to Flight Watch or some other Flight Service Station outlet. Flight Service has a direct link to National Weather Service real-time radar and can suggest a routing that will minimize your exposure to the hazard. Armed with this information, you can return to controller frequency and ask for a reroute.

Many pilots have the luxury of airborne weather radar to aid in storm avoidance. Remember that radar detects not turbulence, but precipitation; a lot of rain or snow might or might not herald a bumpy ride, and dangerous turbulence might exist where no echo shows on radar.

Radar is an active system. This means that the radar set sends out a beam of energy, some of which is reflected back by precipitation to the radar antenna. Some radar radiation continues onward to reflect off the next batch of precipitation, but eventually the beam runs out of strength. This characteristic of being useful for only a certain distance in precipitation is called *attenuation*. The larger a radar dish, or the lighter the precipitation, the less of a factor attenuation is. In most cases, the small radars fitted to light piston airplanes are useful for 5 to 15 miles in moderate rain or snow showers, giving the pilot just enough time to detect and avoid a hazard.

Beware of "black holes" on your radar scope; if surrounded by echoes, a black hole might represent significant precipitation (and the possible accompanying turbulence) that absorbs the radar beam, and not the lack of hazard a blank screen suggests. If you have radar, you probably have an autopilot; this is a good time to enlist "George" because manipulating of radar requires active participation by the operator. Take a good weather radar interpretation course to learn to use this device to its full potential.

Electrical discharge detectors such as Stormscopes and StrikeFinders take a different tack in detecting turbulence. These are passive designs using comparatively little power; as such, they are free from the limitations of attenuation. Electrical discharge detectors also pick up turbulence in the cumulus stage of thunderstorm development, providing earlier warning than radar.

Nonetheless, these devices are not without their limitations, especially in areas of embedded thunderstorms. Electrical detectors assume that lightning strokes are of approximately common power levels and, as such, plot "strong" strokes as closer and "weak" bolts as farther away than they actually are. Embedded storms are generally weaker than freestanding storm cells, so the precise location of cells in areas of stratus is suspect. Add to this the fact that embedded storms typically have fewer lightning events than freestanding cells, giving fewer plots on the detector, and you'll see why electrical discharge detectors are for thunderstorm avoidance, not penetration. Use them to assist you in visually avoiding storms. If coming up on a cloud bank that you suspect might contain thunderstorms, look at the detector screen. If any plots are showing on the screen, avoid the cloud bank altogether. You can't use the plots to accurately navigate around cells once in the clouds.

Dos and Don'ts of thunderstorm avoidance

Here are some general rules about avoiding the hazards of thunderstorms.

Don't:

- Don't land or take off in the vicinity of a thunderstorm. Avoid flight under a thunderstorm or cloud anvil.
- Don't attempt to fly through areas of more than 6/10ths coverage by thunderstorms (to avoid rapidly developing cells).
- Don't fly into areas of embedded thunderstorms without airborne weather radar.

Do:

- Do obtain a good weather briefing and frequent updates once airborne, when flying in conditions conducive to thunderstorm development.
- Do avoid storms by at least 20 nautical miles.
- Do divert or land early to avoid penetrating a storm.

If you can't avoid the storm

If your attempts at avoiding the thunderstorm fail, don't panic. Pilots do successfully fly through thunderstorm cells, generally earning a healthy respect for their avoidance. In order to maximize your chances, then, follow these recommendations.

Before entering the cell:

- Tighten all safety belts and secure loose objects; pick a course (based on the best information you have) to minimize your time in the cell.
- To avoid most icing, pick an altitude below the freezing level, or where the outside air temperature is below –15 degrees Celsius.
- Establish the aircraft's turbulent air penetration speed, and trim for it.
- Turn off the autopilot; it will only fight the air currents and increase airframe stresses.
- Use the tilt on radar, if available, to detect shafts of hail.
- To avoid being blinded by lightning, turn cockpit lights up to maximum.

Penetrating the storm:

- To avoid being blinded by lightning, avoid looking outside.
- Maintain power setting and attitude; accept altitude and airspeed deviations.
- Maintain course to minimize your time in the cell.

Thunderstorms are some of the most awe-inspiring forces on the planet, and they represent a significant hazard to pilots. Know what conditions are conducive to thunderstorm development, ask informed questions when receiving weather briefings, know what to do to avoid the storm, and know what to do if attempts at avoidance fail.

9
Turbulence

RANGING FROM A NUISANCE TO A POTENTIALLY FATAL FORCE, TURBULENCE exists in numerous weather patterns besides thunderstorms. Currently there is no technology in widespread use that can accurately plot the precise location, intensity, or even the existence of all turbulence. Doppler radars, which can track the movement of individual clouds of dust or other airborne particles, are making inroads into turbulence detection, but they do not yet provide coverage over a large part of the continental United States, cannot directly predict future turbulence, and usually are not aligned to detect turbulence aloft. Doppler radars merely detect turbulence near the surface.

Conditions that heighten the probability that turbulence exists, however, are known, so by anticipating the turbulence hazards likely in given conditions, you can plan your flight to minimize turbulence risk. In addition to thunderstorms, there are several other weather patterns that commonly produce turbulence.

CONVECTIVE TURBULENCE

As dissimilar surfaces heat and cool at different rates, columns of ascending and descending air form (Fig. 9-1). Dark surfaces, such as plowed fields, asphalt paving, and deep vegetation, absorb most of the incoming solar radiation and heat the air above them. This creates, as we've seen before, a rising column of air. Lighter surfaces, such

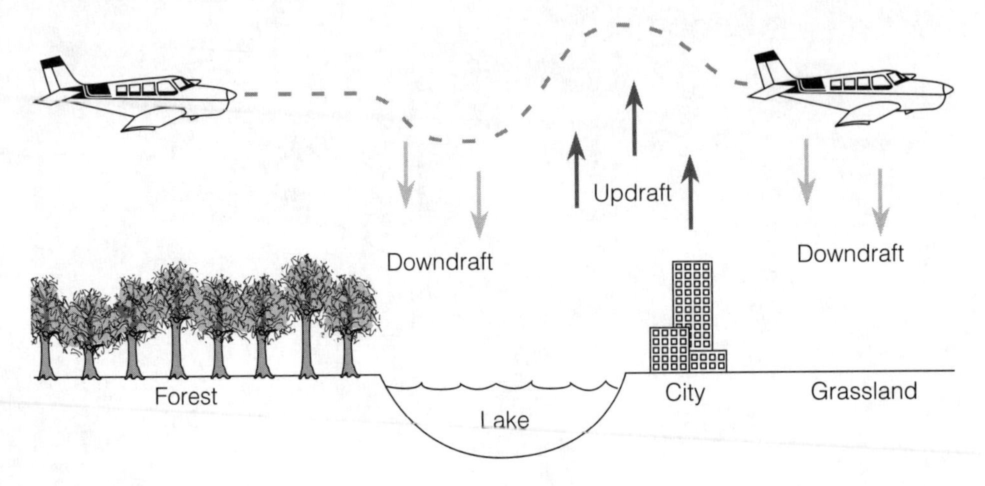

Fig. 9-1. *Convective turbulence over dissimilar surfaces.*

as lakes, rivers, mowed lawns, or thin vegetation, reflect back relatively more radiation; they are cooler than the dark surfaces and consequently form columns of descending air. At the junction of these columns of rising and falling air you'll encounter turbulence. Have you ever flown across a river or a lakeshore on a perfectly still morning? You probably "hit a bump" along the edges of the moister air. Is there a large parking lot under the final approach course? Expect an updraft over that surface, which will tend to make you high on the glidepath, landing long on the runway. Is there cool, wet grass at the arrival end of the runway? Be ready to add a little power to counter a downdraft that tends to make you undershoot the pavement.

Convective turbulence, as it's called, can exist on even a cold winter's day, especially if the sky is clear or scattered clouds allow shafts of sunlight to heat parcels of earth. This bumpiness usually isn't dangerous; instead, it's uncomfortable, especially for passengers that don't fly frequently. Often the vertical extent of the updrafts is marked by a layer of scattered cumulus; these exist like whitecaps on waves of moist, rising air, so expect a bouncy ride beneath or at their level, and a smooth ride up above. Most convective turbulence tops out at less than 5000 feet above ground level, so you can avoid it by staying at higher altitudes as much as possible.

An exception to this is in the desert Southwest, where convection is extreme in the dry, open landscape. Convective turbulence in places such as Arizona and New Mexico can often extend to 15,000 feet above ground level, beyond the capability of most general aviation airplanes to overfly, and it can be intense and dangerous. Most locals in the Southwest will advise you to fly in the mornings and tie down by noon to avoid the turbulence associated with the hottest part of the day.

FRONTAL TURBULENCE

We've already discussed some of the discontinuities, or changes, associated with the passage of a front. Temperature, humidity, and stability might change, but the discontinuity that is always present with a frontal passage is a shift in the direction of the wind.

This wind shift creates a zone of turbulence. The intensity of the turbulence depends on the wind's directional and speed change, and the pressure gradient, or rate of pressure change; this itself is a function of temperature change across the front. Cold fronts tend to involve greater temperature changes than warm fronts, so they are typically more turbulent than warm fronts; their lifting action, wedging warm air upward ahead of the front, adds to the directional shift of wind and increases turbulence. This isn't to say that you won't encounter some turbulence while flying across a warm front; warm frontal turbulence can be intense as well, but cold fronts are characteristically bumpier for pilots.

As altitude increases, again typically, the pressure gradient tends to become shallower (Fig. 9-2). To avoid the worst turbulence of a frontal penetration, then, fly through it at as high an altitude as other conditions (icing, etc.) and your airplane's performance permit. Check the temperatures-aloft forecast and find an altitude with the least dramatic rate of temperature change over distance. Avoid flying along a front, ex-

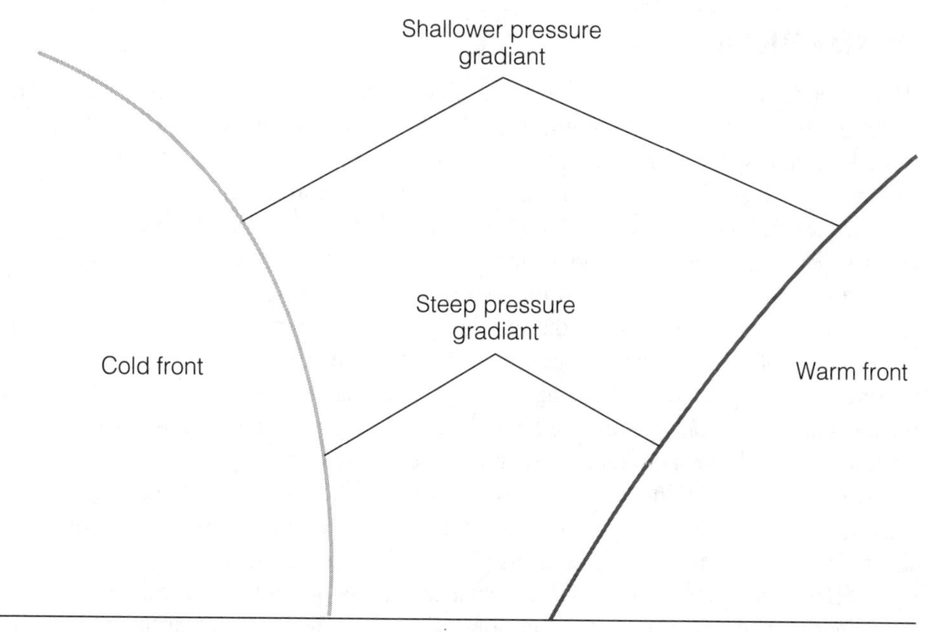

Fig. 9-2. *Frontal turbulence typically decreases with altitude.*

posing yourself to the hazards for an extended period of time; change your flight plan to poke through the front at a right angle, and then fly behind or ahead of the front. You'll have a smoother, safer ride, and you might enjoy clear skies and a tailwind as well (Fig. 9-3).

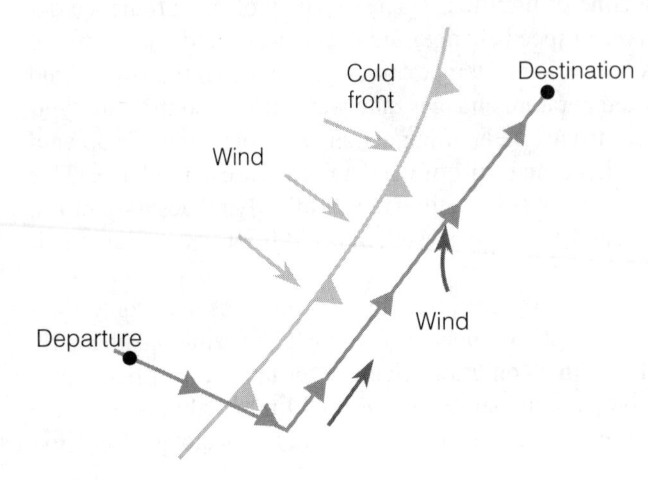

Fig. 9-3. *Plan your flight to minimize exposure to frontal turbulence and to take advantage of wind direction.*

WIND SHEAR

Wind shear is a generic term for any turbulence related to a change in wind speed or direction. In effect, then, all types of turbulence are themselves wind shear. Wind shear has taken on a specific meaning, however; in common parlance the term relates to a change in wind speed or direction that takes place close to the ground.

The aircraft performance effects of penetrating a wind shear zone, which we'll cover in just a moment, are more crucial in heavy, jet airplanes than in lighter, piston-powered airplanes. This is due to the increased weight of the typical turbine airplane (which imparts more inertia and therefore requires a greater distance to reverse the adverse effects of wind shear) and the "spool-up" time required to get full power out of a jet (creating a lag between the onset of a problem and the solution). Lighter, piston planes require less force to reverse a wind shear deflection, and the power necessary is available virtually instantaneously with throttle application.

Because of this, wind shear as an aviation hazard really wasn't investigated until the mid to late 1970s, when heavy, jet airplanes became numerous enough that wind shear accident trends became obvious. Although primarily a hazard to the airliners, nonetheless, wind shear can and does cause accidents in light airplanes.

What sort of performance effect does wind shear have on an airplane? Let's say you're flying in an air mass with a tailwind component, and you fly suddenly into an air mass that gives you a headwind component. With a sudden increase in the relative wind, shearing to a headwind component is a performance-increasing phenomenon.

Indicated airspeed will increase, and altitude will increase, until the pilot compensates and/or the airplane becomes established in the new air mass (Fig. 9-4). Now let's put that tailwind-to-headwind shear on the final-approach course to your home airport. You're on glidepath when suddenly you find yourself both high and fast on final. This is backwards from what you'd normally expect; if you're high, you should have slowed in the climb, and if you're low, your airspeed should have increased. This paradoxical high-and-fast situation tells you that the deviation is the result of wind shear. You're in danger of landing long or even overshooting the runway because of altitude and airspeed. How do you correct for this?

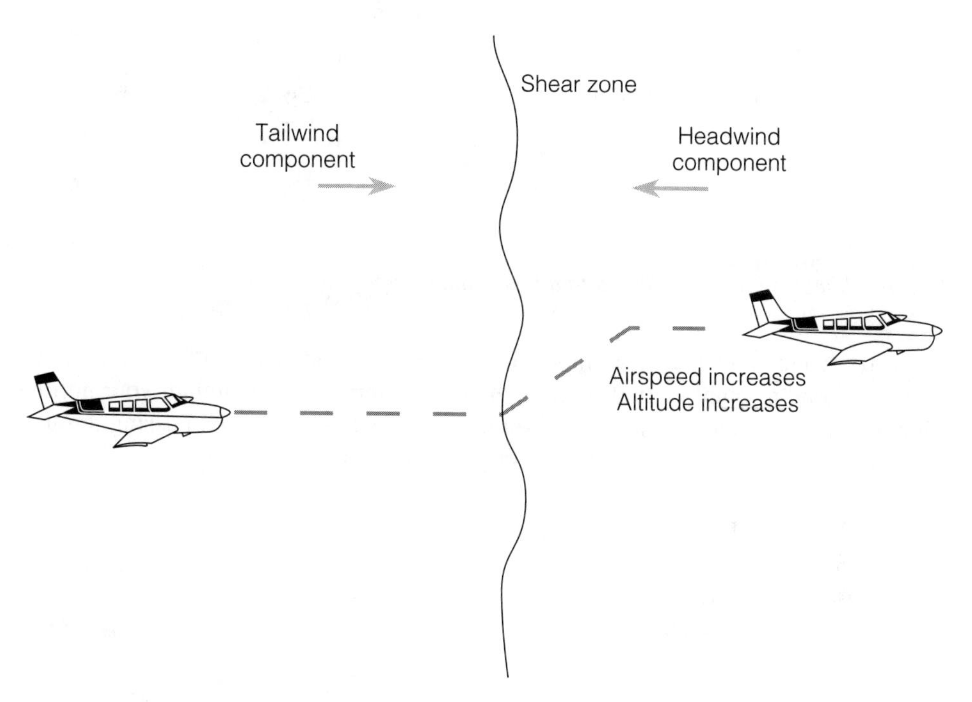

Fig. 9-4. *Shearing from a tailwind to a headwind.*

Make a small power reduction. Pitch the nose down slightly to initiate a trend back towards glidepath. Very soon you'll become established in the new air mass, the increase in relative wind goes away, and your new rate of descent allows a gentle return to glidepath. A sudden nose-down to correct for this wind shear will increase your airspeed further; the airplane will tend to nose up to regain its slower, trimmed speed, and when the relative wind increase goes away, you might be close to a stall. So the answer is a small power reduction (an inch or two of manifold pressure, or a hundred or so propeller rpms in fixed-pitch airplanes) and an accompanying small pitch change until reacquiring glidepath (Fig. 9-5).

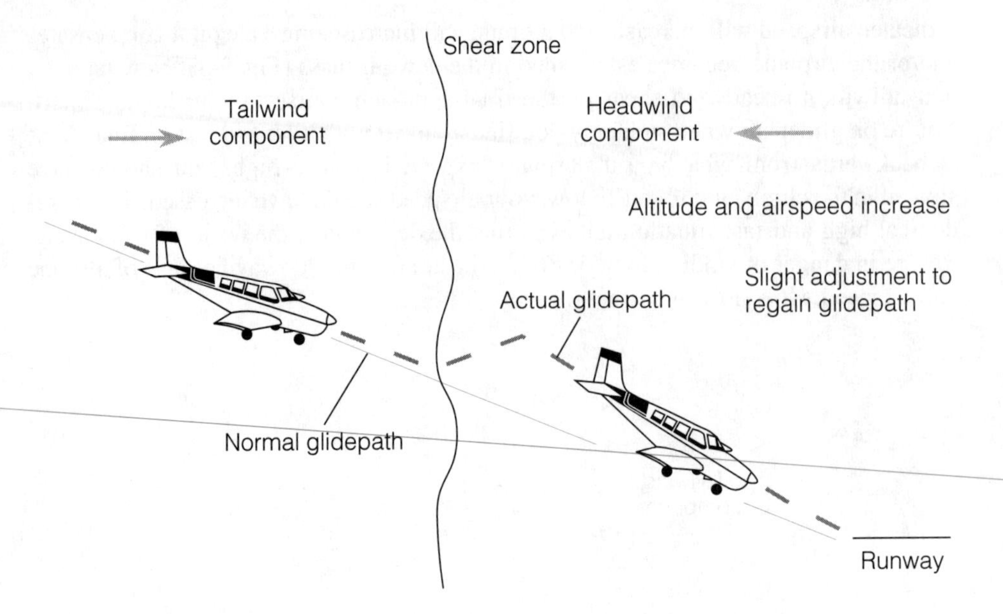

Fig. 9-5. *Shearing from a tailwind to a headwind on approach.*

You encounter a wind shear into an increased headwind on takeoff. Again, the airplane will tend to point up and fly momentarily faster; simply maintain your normal climb pitch attitude, and airspeed will return to normal momentarily. Passing through the bump, you'll end up a little higher than a climbout in still air (Fig. 9-6).

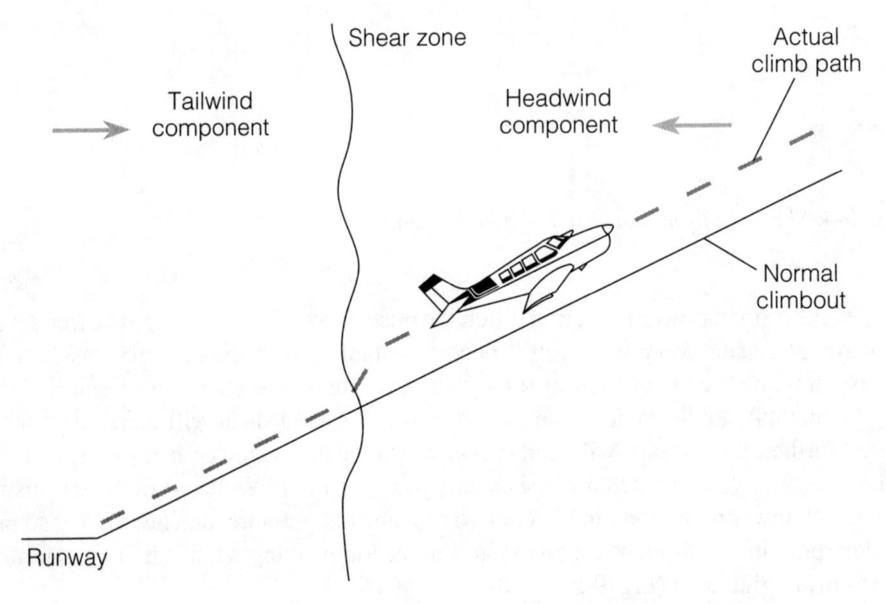

Fig. 9-6. *Shearing from a tailwind to a headwind on takeoff.*

Now we'll look at a wind shear from an area of headwind into an area of a tailwind component. Opposite the shear into a headwind, flying into an increase in tailwind is a performance-decreasing phenomenon. Indicated airspeed will momentarily drop; lift is reduced with the loss of relative wind speed, and the airplane tends to go low and slow on the glidepath (Fig. 9-7). Again, this is the opposite of what you'd expect; if you're slow, you should be high, and if you're low, you should be fast. The unexpected indications should flag "wind shear" in your mind.

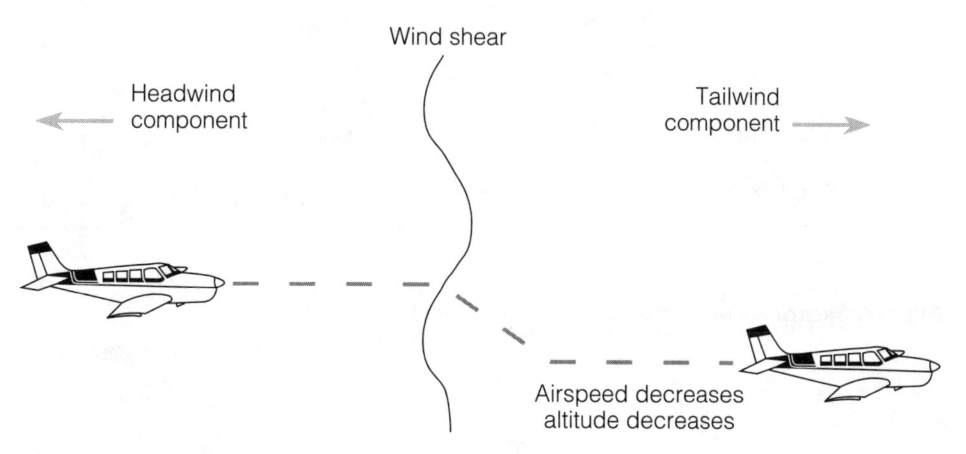

Fig. 9-7. *Shearing from a headwind to a tailwind.*

On final approach, then, you find yourself below glidepath at a less-than-normal airspeed. How do you correct for this? Add power. The airplane will pitch down to regain its trimmed airspeed if you let it, so you'll also have to increase back pressure on the controls to get back on glidepath. A power increase coupled with actively maintaining the correct pitch attitude will create a trend back up towards the glidepath with no tendency to get dangerously slow (Fig. 9-8).

You shear from a headwind into a tailwind component on takeoff (Fig. 9-9). If you let it, the airplane might get dangerously slow while low to the ground; your best bet is to aim for your best angle-of-climb (Vx) pitch attitude and give the engine full power (throttle and propeller) if you aren't there already. You can retract landing gear, as appropriate, if your landing gear doesn't create a drag penalty in transit (check your airplane's *Pilot's Operating Handbook*), but leave flaps where they are. Retracting flaps will increase stalling speed, perhaps enough to cause a stall before you stabilize in the new air mass.

How do you anticipate areas of wind shear near the surface? Watch for high surface wind speeds (15 knots or greater), especially if they're gusty or change direction a lot. Expect wind shear events in the neighborhood of thunderstorms, or for that matter, any precipitation, as rain and snow cool the air, changing the wind flow. Inversions, pockets of warm air overlying cooler air near the surface, often create wind shear, so

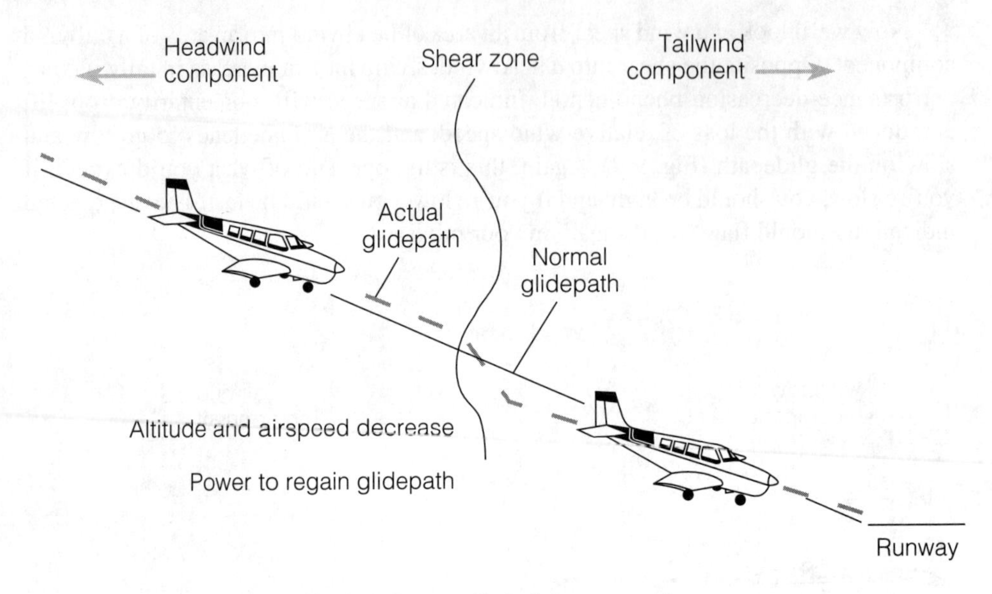

Fig. 9-8. *Shearing from a headwind to a tailwind on approach.*

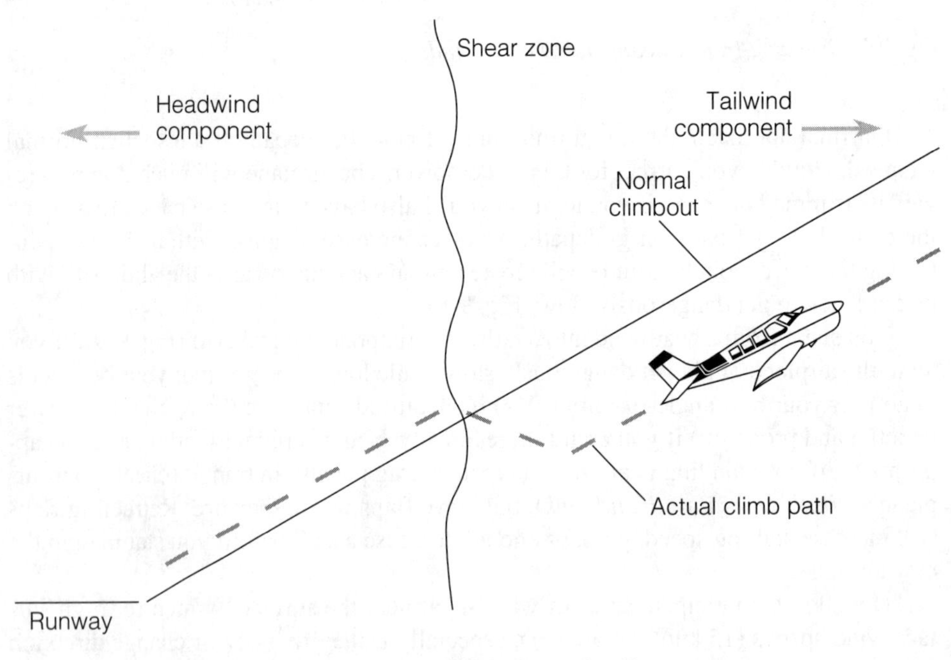

Fig. 9-9. *Shearing from a headwind to a tailwind on takeoff.*

expect the hazard near fronts and at night or on early, clear-morning flights, especially in the summer.

If you anticipate wind shear, what should you do? If your prediction is based on thunderstorms or strong fronts in the area, delay or reroute if possible to avoid the hazard. Otherwise, you can continue, but watch out for the paradox of high-and-fast or low-and-slow on takeoff or landing. Plan on using the longest runway available, and aim for not "the numbers," but several hundred feet from the runway threshold. These precautions will provide you protection from undershooting and overshooting tendencies. Review techniques for correcting for the hazard; remember that flap retraction can be fatal in a wind shear penetration, so don't change wing configuration close to the ground. Finally, don't hesitate to go around or miss the approach if things start to get out of hand. Just apply full power, adjust pitch for the best angle-of-climb, leave flaps where they are, and ride it out.

MICROBURSTS

Microbursts are an extremely strong form of wind shear. They are associated primarily with thunderstorms, especially in the dissipating stage of development, and a phenomenon called virga, rain or snow that evaporates in dry air prior to reaching the ground.

Any descending column of air can create a microburst. There are two types of microbursts: "wet" and "dry." A *wet microburst* is one where moisture is visible in the outline of the downdraft, such as in a rain shaft near a storm cloud. A *dry microburst* is formed in arid regions, beneath storm clouds or virga; evaporation is a cooling process, and the evaporation of precipitation further cools the air, increasing its tendency to descend. Microbursts of varying intensity are almost always present when virga is seen.

What does a microburst do? This rapidly descending column of air strikes the surface, spreading out like ripples on a still pond. Descending air above compresses the airflow, keeping it close to the ground; this creates a natural venturi effect that actually accelerates the wind speeds. Speeds of 90 knots in microbursts are common; the friction zone, between relatively still air and this fast microburst flow, creates intense turbulence. At its edges, slowing microburst winds curl upward, creating more turbulence; in cross-section, a microburst looks something like a canoe (Fig. 9-10).

This accelerated wind flow can be extremely dangerous; in fact, they're often mistaken for tornadoes when they hit invisibly after dark. I was sitting at home late one evening, watching the local news network warn of severe weather passing overhead, when suddenly the house began to shake and I heard branches snapping off the trees outside. I was just getting up to get my wife and child into the basement when the winds ended; I wasn't certain whether a tornado had passed or not. The next morning, as I drove to work, trees, lawn chairs and garbage were strewn everywhere on my block and the ones around it; elsewhere, it was as if nothing had happened at all. I assume my neighborhood was hit by a microburst, triggered by a severe thunderstorm overhead; I would hate to have been trying to fly through that wind.

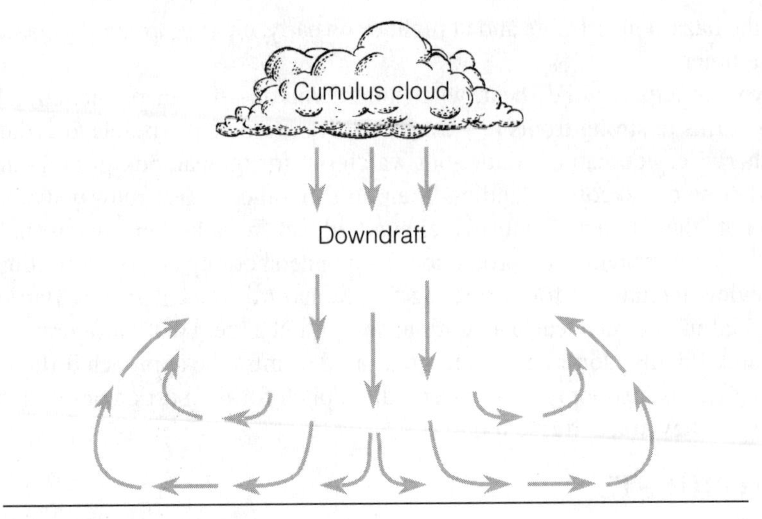

Fig. 9-10. *Cross-section of a microburst.*

Microbursts have a life cycle lasting as much as half an hour. They tend to build in intensity in the first five minutes and then gradually dissipate as they blow with the prevailing winds across the ground. Because microbursts often form in the final stage of a thundercloud's life and can last for some time afterward, you might encounter a microburst when there's no visible clue to its formation. In a moment, we'll look at times to anticipate microburst formation.

Don't rule out a microburst in the early stages of a storm, or away from heavy precipitation, when virga is likely. In fact, microbursts might form when the producing cloud has insufficient moisture to appear on radar.

In the mid-1980s, the FAA, NASA, and the National Weather Service conducted a 90-day microburst study on the grounds of Denver's Stapleton Airport. During that three-month period they recorded 160 microbursts. Only 5 percent of those occurred when a thunderstorm was overhead; 50 percent were associated with virga in sight of the field, and 50 percent happened when only weak or even no convective activity was visible from the airport grounds. Significantly, almost all microbursts formed between 1400 and 1800 local.

Although I've already said that heavy jets are more susceptible to wind-shear accidents than light airplanes, the NTSB assigned microbursts as the primary cause of 91 general aviation accidents in the years 1983–1990. The majority of these happened in Colorado and California. Microbursts are almost certainly quite common in the arid states of New Mexico, Arizona, Nevada, and Utah as well, but microburst-related accidents are less numerous simply because of the relatively few number of flights in those areas.

So how do you survive a microburst? How can you anticipate their formation? What indications result from a microburst penetration? The best way to survive a microburst is to avoid it. Anticipate the possibility of microbursts if one or more of these conditions prevail:

- Surface temperatures exceed 80 degrees Fahrenheit. This heat allows contrast with rain-cooled air to create downdrafts.
- Temperature/dew-point spreads exceed 30 degrees Fahrenheit. This sets the stage for virga, almost always a microburst producer.
- ATIS updates, wind checks from the tower, or shifting wind socks. Rapid changes might be the result of microbursts.
- Low-level wind shear alerts from ATC.
- Pilot reports of airspeed gain or loss on approach or landing. Airline pilots are usually pretty good about making these reports.
- Thunderstorms, virga, or towering cumulus clouds visible, especially downwind of your location. Microbursts might blow with the prevailing wind towards your flight path.
- Blowing dust, dust devils, waving grasses, or other indications of strong winds at the surface. These might point out the boundaries of microbursts.

If any of these cause you to anticipate microbursts, what should you do?

- Divert or delay, if possible, to avoid the threat. Change arrival or departure time to miss the prime 1400–1800 formation time.
- Use the longest runway available. This gives you room for a wind-shear-related undershoot or overshoot.
- Plan to maintain flap configuration until well aloft, to minimize the chances of a turbulence-induced stall.
- Watch for airspeed loss, increase, or hesitation on takeoff and landing. Airspeed, a performance measurement, is the result of power and pitch attitude. If you have power and pitch set, unexpected airspeed behavior might indicate a microburst penetration.
- Use maximum takeoff power, if flying an airplane approved for reduced-power takeoffs.
- On landing, fly final approach (Vref) speed plus 10 to 20 knots, providing a cushion against a turbulence-induced stall.

You anticipate the microburst and you adjust your technique to look for and guard against penetration indications. But what will those indications be? How will you know you've penetrated a microburst?

Entering a microburst from most angles creates a strong shear to a headwind, so watch for the high-and-fast indication (Fig. 9-11). Correcting like you would for "standard" wind shear, however, might set you up for a fatal descent. After reducing power, lowering pitch, and increasing your rate of descent, you might enter the downdraft cen-

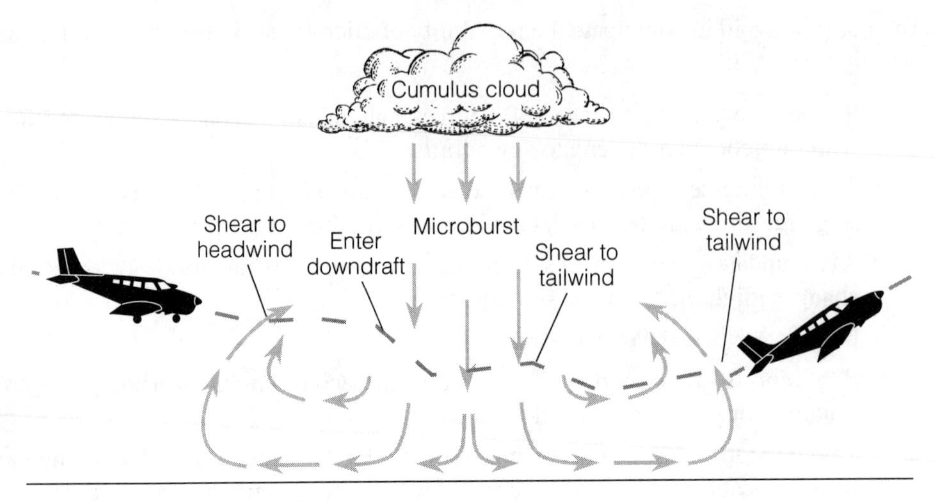

Fig. 9-11. *Penetrating and escaping a microburst.*

ter of the microburst. Airspeed will hesitate, and vertical speed will increase. Now you power and pitch up to climb out of the rapid descent, only to encounter the shear to tailwind associated with the outflow of the event. If you're not careful, the pitch attitude you established for climbout might cause a stall as relative wind flow decreases. The oft-discussed Lockheed L1011 microburst accident at Dallas in the early 1980s ended with a full-power, nose-up collision with the earth.

How, then, do you survive entering a microburst? Upon noticing the first indication of penetration, apply maximum power and pitch for your airplane's Vx speed. Retract landing gear if appropriate, but leave flaps where they are until reaching at least traffic pattern altitude at a safe airspeed. Accept the reduced indicated airspeed you register at the beginning of the go-around. If landing, remember that microbursts will blow downwind, so a short delay followed by another attempt might keep you away from harm, but also remember that microbursts are a group phenomenon; watch for the indications again on subsequent landing attempts. Whether taking off or landing, file a pilot report when time permits to help the next pilot make a go/no-go decision.

CLEAR-AIR TURBULENCE

Clear-air turbulence is another generic term, this one defined as turbulence with no obvious, visible cause. It has come to mean primarily, however, turbulence associated with the jet stream. We'll define clear-air turbulence, or CAT, as it's commonly known, as such; we'll avoid CAT by avoiding flight near or under the jet stream.

We've already discussed the jet stream as a high-speed river of air that acts as a steering current for moisture over the continent. If you were to look at a commercial weather presentation, like The Weather Channel or a newspaper, you'd see the jet stream depicted as a continuous river of air, meandering across the states from coast to coast. This depiction does indeed show the general flow of the atmosphere, but it does not accurately show the well-developed jet "cores" that create flying hazards.

A little review about the jet stream: the jet forms at the tropopause height, the boundary between the troposphere and the stratosphere. This height is usually around 35,000 to 50,000 feet in summer over the continental United States, and the height can drop to half that value in the winter months. Wind speeds in the jet are by definition greater than 60 knots, but hazards don't really begin to arise until wind speed exceeds twice that value. If the wind flow in the jet core is greater than 150 knots, severe turbulence can form as much as 20,000 feet beneath the tropopause height, affecting even the lowest-flying aircraft in winter. The turbulence typically extends 50 to 100 miles on the north, polar side of the core as well, so avoiding areas beneath and north of well-developed jet cores is necessary to avoid encountering CAT (Fig. 9-12). We'll look shortly at how to detect these jet-core locations using weather charts.

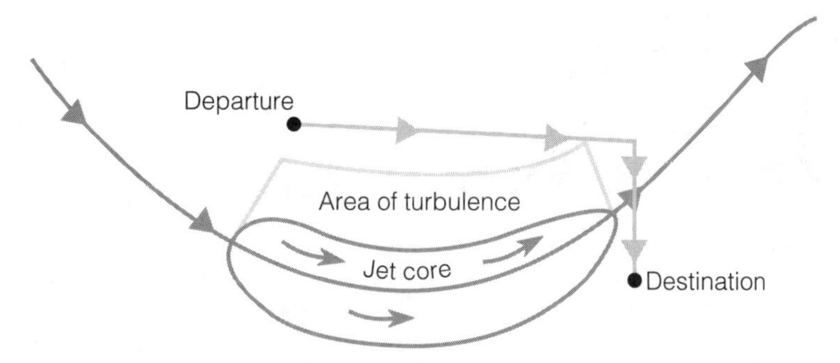

Fig. 9-12. *Plan your flight to avoid probable areas of clear-air turbulence.*

MOUNTAIN WAVE

Mountain wave turbulence is created by high wind speeds, generally 30 knots or greater at the ridge height, which flow roughly perpendicular to ridge lines in stable air. Unstable atmospheres tend to break down the natural laminar flow that creates the mountain wave; the phenomenon is most common in winter (when jet-speed winds are closer to the surface) and tends to break down at night, even if the winds don't subside.

How does the wave form? This wind is deflected upward by the mountain; as it rises it also cools, until eventually it becomes cold enough that it begins to descend. On the way down it heats, until either it rebounds off the surface or warms to the point that it begins to rise again. Waves of ascending and descending air can extend several hundred miles downwind from the peaks.

Air inside the mountain wave is quite smooth. Sailplane pilots love to ride its lifting action to set new endurance records. Around the boundaries of the wave, however, turbulence can be extreme.

Lenticular clouds might mark the location of a mountain wave. Lenticular, or lens-shaped clouds, appear to hold position over the earth while clouds at other altitudes zip by in the high winds. This is because the standing lenticular, as it's often called, is constantly forming and evaporating as moisture flows through the wave. Water in vapor form condenses as it cools in the upper part of the wave, forming a visible cloud, but revaporizes when warming on the way back down. Moisture, then, is constantly added and removed from the cloud; hence, the lenticular appears to remain in one place. Moisture might also condense in rotor clouds near the surface, where rebounding creates a cooling venturi effect (Fig. 9-13).

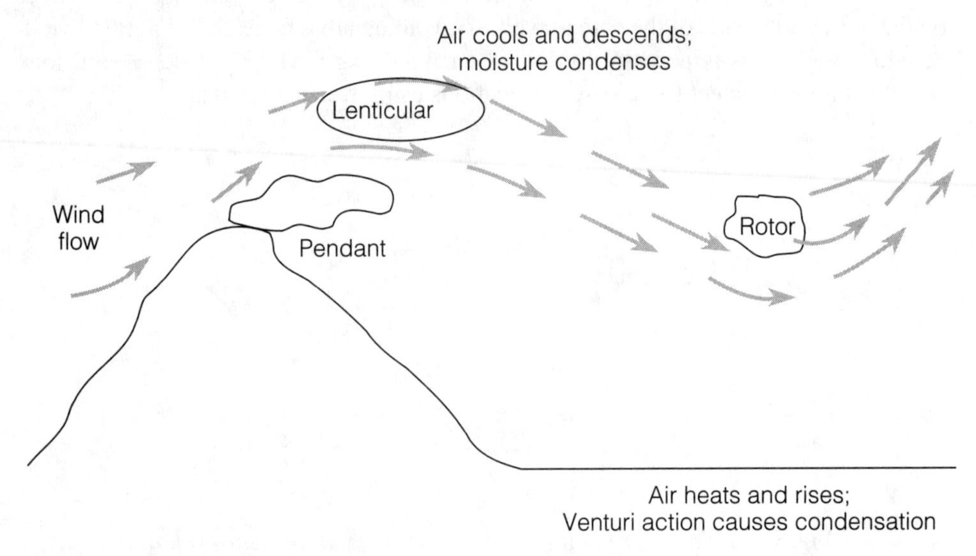

Fig. 9-13. *Formation of a mountain wave.*

How do you avoid mountain-wave turbulence? Avoid flight near and downwind of mountains if these forming conditions exist:

- Wind at the ridge height is roughly perpendicular to the ridge.
- That wind speed is 30 knots or greater.
- The air mass is stable.

If you can't avoid these conditions, you can route yourself as far as possible from the ridge lines, between lenticulars (in the still air between the waves), or fly at night, when mountain waves tend to break down. Fly on the upwind side of the ridge as much as possible, in the zone of rising, not descending, air. Beware of very smooth air near the mountains; this points to a stable air mass, one element of mountain wave formation. If you do penetrate the mountain wave, slow down and pick an escape route away from the downwind side of the peaks. Avoiding mountain-wave conditions and locations won't assure you a smooth ride near the peaks; on the contrary, wind flows around hills like water across pebbles, turbulent and unpredictable. You'll just avoid the extreme turbulence associated with mountain waves.

WEATHER PRODUCTS THAT WARN OF TURBULENCE

Like thunderstorms, there are several types of aviation weather products that warn of turbulence. In fact, by definition, any warning of thunderstorms implies moderate or greater turbulence, so all thunderstorm-related charts also warn of strong turbulence. Now we'll look at ways to anticipate nonthunderstorm-related turbulence.

Surface aviation reports and winds aloft

The "hourlies," observations at reporting point, can warn of turbulence (Fig. 9-14). If the surface wind is greater than about 15 knots, expect light turbulence near the ground. If surface winds exceed 25 knots, or you see a "G," for gust, in the wind portion of the report, plan for moderate or greater turbulence in the first few thousand feet above ground level.

```
3KM RS 1745 E100BKN 20 OVC 10 25/M/3220G40/006
```

Fig. 9-14. *A surface aviation weather report indicating likely low-level turbulence.*

You can also compare the wind direction and speed from a surface report to the lowest level above that station on the winds aloft forecast (Fig. 9-15). If the wind speed more than doubles, and/or the wind direction changes more than 40 degrees, in the lowest level of the atmosphere, expect light to moderate turbulence along the shear. This seems to happen most commonly on cool, clear nights when rapid radiational cooling near the ground alters normal wind patterns.

Look at the wind speed and direction for altitudes above a single reporting point. If the pattern differs from the typical, gradual changes towards the southwest-to-northwest, surface-to-60-knots at the tropopause pattern, expect turbulence. If nearby reporting points indicate a big shift in wind direction or speed for a given altitude,

```
MEM SA 1752 M27 BKN 250 OVC 20 108/50/39/2715/985

              FD WINDS ALOFT FORECAST

              3000          6000
MEM           2132          2540+02
```

Fig. 9-15. *Compare wind speed and directions from a surface aviation report (top) to those at the lowest reported altitude on the winds aloft forecast (bottom) to predict low-level turbulence.*

anticipate a bumpy ride as well. The more pronounced the discontinuity, the more likely, and more intense, turbulence will be.

Stability chart

We've already discussed the lifted index in the previous chapter. If a negative number exists along your route of flight, expect a bumpy ride, even if thunderstorms don't form. Typically, positive numbers indicate smooth flying conditions, but beware of stable air, essential to the formation of mountain waves, in areas downwind of ridges.

You'll want to reference the stability chart in winter primarily to detect mountain waves "out West" and on the east side of the Appalachian chain (Fig. 9-16).

Tropopause height/vertical wind shear

The tropopause height/vertical wind shear chart is not widely known, but it provides a wealth of information about the location and altitudes of potential clear-air turbulence (Fig. 9-17). Derived by the same upper-level observations that create the stability chart, and therefore recorded at 00Z and 12Z daily, this chart gives you two bits of useful information: the altitude of the tropopause and the location of developed jet stream cores.

Lines across the chart indicate the height of the tropopause expressed in terms of flight level. The dashed lines marked "2K," "4K," and "6K" indicate the rate at which wind speed changes with an increase in altitude. For instance, "6K" means that the wind velocity varies by 6 knots for every 1000 feet in altitude change in that area. Values of 6K or greater indicate a jet stream; expect clear-air turbulence in that area from the indicated tropopause height to as much as 20,000 feet below that altitude, and extending 50 to 100 miles on the north, cold side of the jet core.

Pilot reports, area forecasts, and terminal forecasts

Pilot reports are the only reliable means of determining the presence or absence of turbulence. When evaluating a pilot report, ask yourself what the probable experience level of the issuing pilot might be (based on aircraft type and altitude); inexperienced pilots tend to overstate turbulence, while high-timers tend to downplay its intensity. We'll look at the specific reporting criteria later in this chapter.

Area forecasts include a statement about probable locations of turbulence, as well as its intensity. You can use terminal forecasts in the same manner as surface aviation reports to predict the likelihood of turbulence near the ground (Fig. 9-18).

Low-level significant weather prognostic

The "prog charts" provide a forecast of widespread areas of turbulence, as well as their expected altitudes (Fig. 9-19). A single "hat" symbol indicates light to moderate turbulence, while a double "hat" warns of severe or extreme instability.

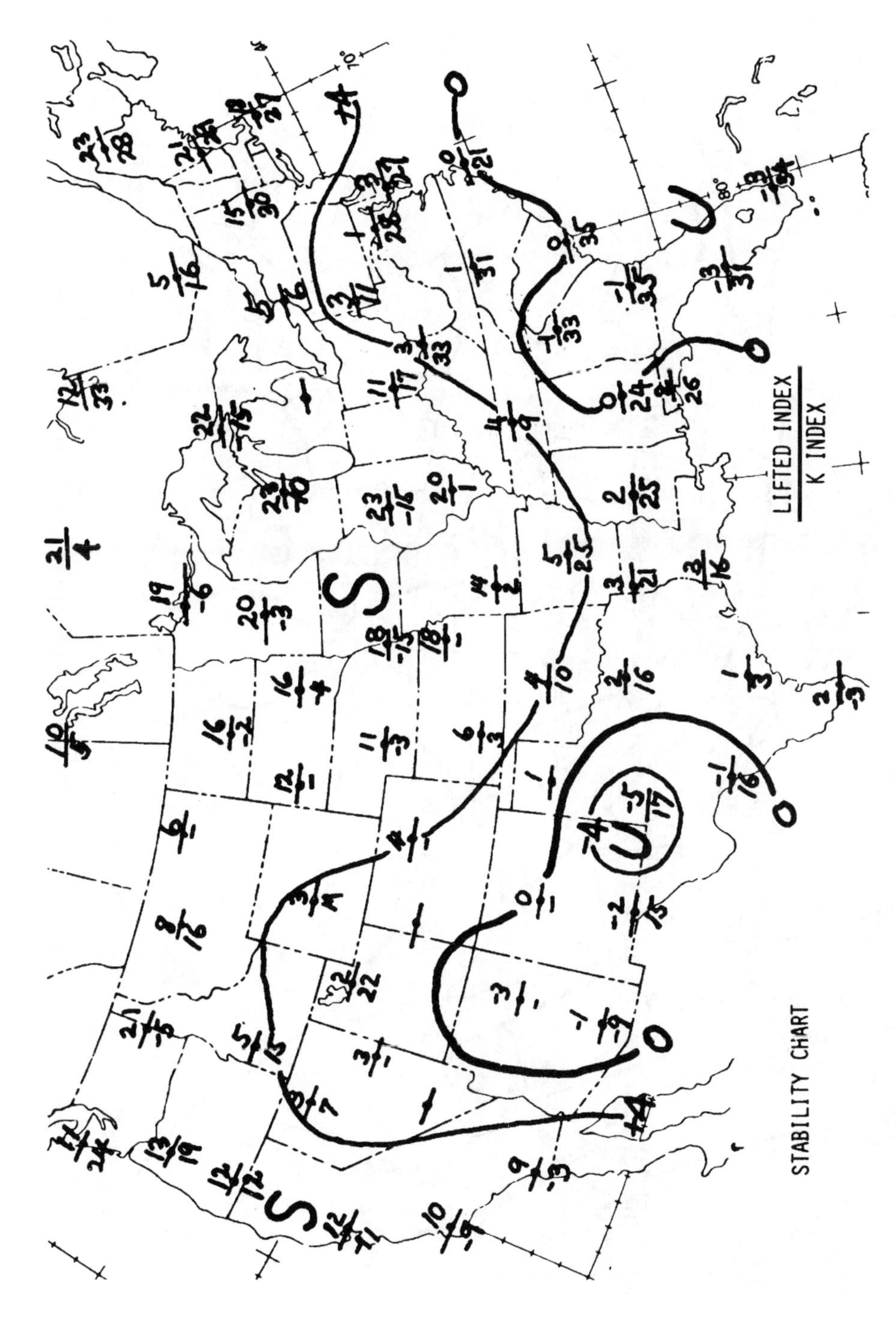

Fig. 9-16. *A stability chart (Advisory Circular 0045B).*

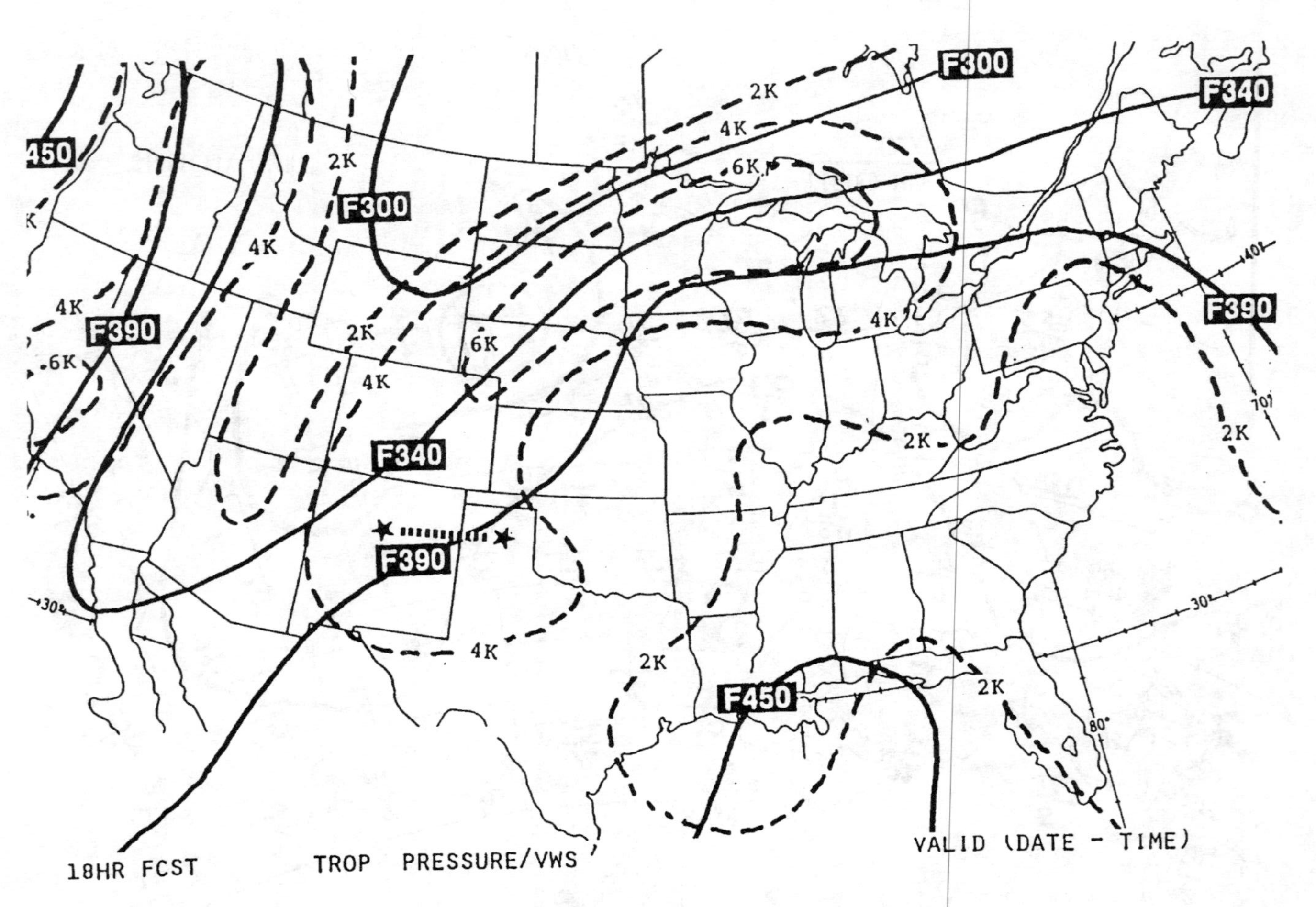

Fig. 9-17. *A tropopause height/vertical wind shear chart (Advisory Circular 0045B).*

```
MEM UA/OV MEM/TM1750/FL090/TP BE02/SK CLR/TA-59/WV241099/TB
LGT


FA 231045
GA
NRN AND CNTL GA..15-25 OVC LYRD TO 180. VSBYS 3-5R-F. SCT
TRW..WITH LNG/CLUSTERS EMBDD TRW PSBL. TSTMS PSBLY SVR. CB
TOPS TO 400. OTLOK..IFR CIG R F.


ICT FT 231818 15 SCT C25 OVC 3416G26 OCNL C15 OVC 4S-.
    20Z VFR WND.
```

Fig. 9-18. *A pilot report (top), an area forecast (center), and a terminal forecast (bottom), all indicating turbulence.*

AIRMETs and SIGMETs

Airman's Meteorological Information, or AIRMETs, point to hazards that threaten primarily light aircraft. AIRMETs are issued as needed for areas of moderate turbulence or surface winds exceeding 30 knots. All AIRMETs pertaining to turbulence have the identifier "Tango" with a number in order of issuance ("Tango 1," for instance).

A SIGMET, or Significant Meteorological Information report, will be issued for existing or potential severe or extreme turbulence. They are identified by a letter (November through Romeo, and Uniform through X-ray), as well as a number in order of issuance. On a daily basis, once a letter has been assigned to a specific type of hazard ("Romeo 1 for turbulence," for example), all subsequent reports will bear the same alphanumeric (Romeo 2, -3, etc.) (Fig. 9-20).

AIRMETs and SIGMETs are broadcast over navigational aids and ATC frequencies on issuance, and then at 15 and 45 minutes after each hour as long as they are valid.

Turbulence reports

All along I've alluded to different intensities of turbulence. Now we'll define those levels of turbulence and point out some assumptions that Flight Service makes when predicting the possibility and intensity of turbulence.

Turbulence is categorized by the effect it has on airplanes flying through it. Unfortunately, an airplane's mass, airspeed, and wing loading are going to change the per-

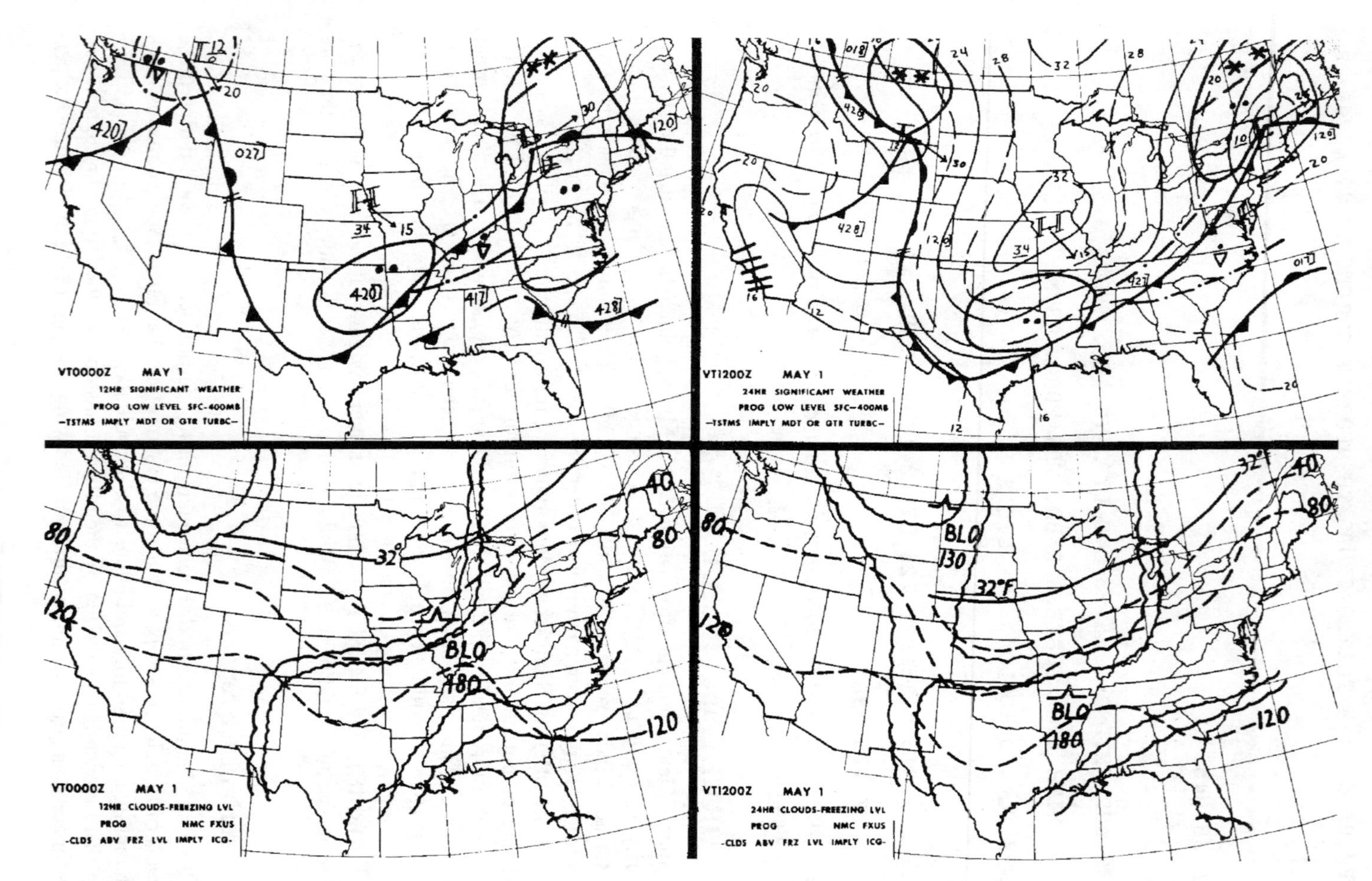

Fig. 9-19. *A low-level significant weather prognostic (Advisory Circular 0045B).*

```
AIRMET TANGO FOR TURBC..LLWS AND STG SFC WNDS VALID UNTIL
18200
AIRMET TURBC..SD NE KS OK TX
FROM 50SW DIK TO OMA TO DFW TO 40W MAF TO DHT TO GLD TO BFF
TO 50SW DIK
OCNL MDT TURBC SFC-160 DUE TO STG SUF WNDS. CONDS CONTG BYD
20Z.


SIGMET NOVEMBER 1 VALID UNTIL 182200
SD NE KS IA MO OK
FROM FSD TO DBQ TO COU TO SGF TO TUL TO GAG TO ONL TO FSD
OCNL SVR TURBC BLO 50 DUE TO STG SLY LOW LVL FLOW. ACFT RPTD
MDT-SVR TURBC BLO 30 NR MKC AT 1720Z. CONDS CONTG BYD 22Z.
```

Fig. 9-20. *An AIRMET (top), and a SIGMET (bottom), calling for turbulence.*

ceived intensity of turbulence, hence, consider turbulence reports skeptically; the effect you experience might differ from the reports.

In order to standardize turbulence forecasts, then, the FAA has chosen a "typical" airplane type when forecasting the effect of turbulence. According to the Aviation Unit of the Severe Storms Forecasting Center in Kansas City, the agency responsible for area forecasts as well as in-flight advisories, this "typical" airplane used in estimating bumpiness is a Saab 340 commuter twin. This falls into line with the "airline-style" definitions used to characterize turbulence; remember that a gust that imparts "light" turbulence on a Saab might well create "moderate" effects on the Cessna you're flying. These are the definitions of turbulence intensity used in FAA/NWS forecasts and reports:

Light turbulence causes slight, erratic changes in attitude and altitude, and slight strain against seat belts; unsecured objects are displaced slightly. Food service is okay; walking is possible without difficulty.

Moderate turbulence is of greater intensity, but the aircraft pilot is still in control at all times. There is strain against seat belts; unsecured objects are displaced. Food service and walking are difficult.

Severe turbulence causes large, abrupt attitude excursions. The aircraft is momentarily out of control. There are large variations in indicated airspeed; occupants are forced violently against seat belts. Unsecured objects are thrown about; food service and walking are impossible.

Extreme turbulence tosses aircraft violently about and is virtually impossible to control. It might cause structural damage.

Sometimes you'll hear reports of the frequency of turbulence. *Occasional turbulence* is that which occurs less than ⅓ of the time. *Intermittent turbulence* happens ⅓ to ⅔ of the time; *continuous turbulence* is evident more than ⅔ of the time. *Chop* is a rhythmic, repetitive turbulence of any intensity and/or frequency; you might hear of "occasional light chop."

Currently there is no technology in widespread use that can detect the existence of turbulence. The only positive indication of the existence (or absence) of turbulence is a pilot report. When giving a PIREP of turbulence, remember to accurately state the intensity based on what indications you experience. How many times have you reported severe turbulence? Was the airplane really momentarily out of control, or were you just very uncomfortable? An accurate pilot report is essential to the next pilot's go/no-go decision; you depend on accuracy when planning your flight as well.

When evaluating a pilot report of turbulence you hear over the radio or obtain from a weather outlet, ask yourself two questions. First, what type of airplane made the report? If the airplane is similar in weight, size, and speed to the one you're flying, the turbulence intensity should be the same for you, assuming an accurate report. Was moderate turbulence reported by a King Air? It might affect your Piper more intensely.

Second, if possible, make a subjective judgment about the probable experience level of the pilot making the report. Of course this is vast generalization, but you can usually expect a PIREP from a light, single-engine airplane flying low to have been made by a relatively low-time pilot; such pilot reports tend to overstate intensities. Reports from planes generally flown by professionals, such as corporate jets and airliners, tend to be more accurate, although sometimes "high-timers" tend to downplay turbulence's severity.

Where turbulence is likely. Weather phenomena and terrain provide clues to the possibility of turbulence. Although the presence or absence of turbulence is never assured, and its intensity might vary, you can generally expect turbulence of stated levels in these situations:

Light turbulence. Light turbulence is common near large hills or mountains when winds are light. It forms in clear-air convective currents such as over the boundaries of lakes and fields on a sunny day, and it is sometimes capped by fair-weather cumulus clouds. Light turbulence is also common beneath troughs and lows aloft and near the surface under a high-flying jet stream. Expect light turbulence within 5000 feet of the ground when winds near 15 knots at the surface, or in an inversion like you'd find in early morning after a clear, cool night.

Moderate turbulence. Anticipate moderate turbulence near mountains with a 15–50 knot wind, especially in a mountain wave. You'll find this level of intensity in towering cumulus and near dissipating thunderstorms. Turbulence encountered flying through a front is usually moderate, as is the instability associated with the jet stream. Moderate turbulence is common near the ground when surface winds exceed 25 knots, or strong surface heating is present, such as in the desert Southwest.

Severe turbulence. If a mountain wave forms in winds exceeding 50 knots at the ridge height, or a jet stream core surpasses 150 knots wind speed, the associated areas of turbulence are likely to be severe. Severe turbulence is also common in mature thunderstorms.

Extreme turbulence. Extreme turbulence is usually evident in rotor clouds associated with a mountain wave, and in and around severe thunderstorms and microbursts.

Turbulence might be merely uncomfortable, or it might be intense enough to damage or even destroy your airplane. Know when to expect turbulence, and, when obtaining a weather briefing, ask questions in order to predict the likely locations and strength of turbulence. Have a plan in mind for dealing with anticipated turbulence, and know how best to escape if the turbulence becomes hazardous.

10
Reduced visibility

AS AN AVIATION WEATHER HAZARD, REDUCED VISIBILITY CONTRIBUTES TO more fatal accidents than any other. How many times have you read of "continued VFR flight into instrument conditions," or "pilot descended below altitude minimums in poor visibility?" Knowing something about the causes of reduced visibility, and when it is likely to occur, will help you make a safer go/no-go decision before takeoff, and to reevaluate that decision in the air before it's too late.

There are several causes of reduced visibility. With the exception of dust storms, blowing snow, and smoke from fires and volcanoes, all problems of reduced visibility are related to condensation. We've already looked at the general rules governing condensation, but a quick review is in order. Condensation occurs whenever the temperature/dew-point spread nears zero, or when the relative humidity is 100 percent. The intensity of the resulting visibility loss depends on the availability of condensation nuclei; places with large levels of airborne pollution, such as the Northeast, have typically lower visibilities than areas with less polluted air, like the Great Plains.

Condensation is most likely to take place in the coolest part of the day, usually at or shortly after dawn. If the temperature/dew-point spread is three degrees at 4 a.m., for instance, expect conditions to get worse; if the spread is three degrees at 9 a.m., typically the visibility will get better. Now we'll cover several causes of reduced visibility: low clouds, fog, mist and haze, precipitation, blowing snow, sand or dust storms, and smoke.

LOW CLOUDS

Clouds will form at whatever altitude the temperature and dew point meet. If this altitude is low enough, the clouds might well present a hazard to pilots. Because of the differing temperature lapse rates of moist versus dry air, it's possible to derive an estimation of the altitude of the bases. Typically, if you take the difference between the temperature and dew point, in degrees Fahrenheit, and multiply that by 230, you'll have the approximate cloud base. A three-degree temperature/dew-point spread, then, reveals a cloud base near 700 feet. Strong winds might break this layer up or cause it to ride higher. From a practical standpoint, if you have the dew point information, you usually have the cloud base report as well, but if you're flying to a coastal area not served by a weather reporting station, you can guess that the greater amount of water available means the cloud bases will be lower. Flying to St. Augustine, on the Florida coast? Expect the clouds to be lower than inland reporting points like Orlando.

The FAA and National Weather Service have set definitions for the extent of cloud coverage. A *clear sky* is where less than 10 percent of the sky is clouded. A report of *scattered clouds* means from 10 to 50 percent of the sky is covered; a *broken layer* hides 60 to 90 percent of the sky. Anything more than 90 percent coverage is deemed an *overcast.*

A cloud *ceiling* is the lowest level of clouds that contributes to a broken or overcast layer. An exception is when that layer is designated *thin*, which means the sun or moon is visible through the clouds; such a cloud cover is not called a ceiling. Ceilings are reported as either *measured* (determined by triangulation, a timed balloon launch, or a light-beam rangefinder) or *estimated* (a subjective judgment made by a trained, ground-based observer). Pilot reports are also considered estimated ceilings.

You'll sometimes hear reports of an *obscured* sky or a *partial obscuration.* Any obscuration, whether it be partial (10–90 percent of the sky) or full, means that some surface-based phenomenon prohibits the observer from seeing the cloud bases. ATIS might warn of "ceiling 800 overcast, partially obscured, fog." This tells you that you might or might not see the ground out of the instrument approach because of fog. What's the temperature/dew-point trend? Is the air cooling or warming? If you know, you can decide whether it's worth holding for half an hour for things to clear, or if you need to divert to an alternate. Is the obscuration caused by blowing snow, for instance? Unless a frontal passage is imminent, which will change the wind flow, forget about landing there anytime soon. Is precipitation the culprit? Ask for a radar update of conditions upwind. The rain or snow might end soon, removing the obscuration and allowing you to make the approach and landing successfully after a short wait.

FOG

If there's enough moisture present that the temperature and dew point meet at the surface, the resulting cloud is called a fog. Fog has properties of uniform density and a definite boundary when viewed from the outside; it's a cloud that rests on the ground. Flight Service and the National Weather Service will call fog "thick" if the visibility is less than 1000 feet.

Fog can form quite quickly. It takes only a small temperature change to hit the saturation point; a small increase in wind from a wet source can cause condensation, and even dust picked up into the air can add condensation nuclei to the mix, lowering visibility. I once worked with a man who flew Douglas C-124s across the Atlantic (back when big airplanes were called four-motor transports). He told me that it was common in Greenland for the wake turbulence of a Globemaster takeoff to drop visibilities below instrument minimums for up to half an hour in the cold, moist environment. I had a similar experience instructing in a Bonanza. The ceiling was reported right at minimums for the ILS approach into Wichita; every time we practiced the approach behind an airliner, we had to miss at decision height; otherwise, we broke out and could make a landing. If you have to miss an approach others are completing, and you are "number two" behind a heavy transport, you might be justified in trying the approach a second time. But avoid the temptation to go "just a little lower;" you don't want to be someone a pilot thinks about when he reads the first paragraph of this chapter!

There are several different types of fog: radiation, or steam fog; advection fog; and upslope fog. Let's look at how they form:

Radiation or steam fog. This is the sort of visibility limiter you find early in the morning after cool, clear nights. Radiational cooling has lowered the air temperature near the surface to the saturation point; the resulting inversion traps the moisture in, creating a fog. Winds have to be light; otherwise, the fog is dispersed or blows up to become a low cloud layer. This fog is sometimes localized, such as over a small pond or a river, or it can be extensive, fogging in areas as large as several states. Watch the temperature/dew-point trend to anticipate radiation fog; plan your departure for before dawn or late morning to avoid the hazard.

Advection fog. Advection fog is created when moist air blows over a cooler surface. The West Coast of the United States, as we'll see in the chapter covering regional weather patterns, is famous for this scenario. You'll also see it in the Great Lakes region in the winter; air is warmed as it crosses open water, collecting water vapor along the way; when that wet air begins to cool on contact with cold earth downwind of the lake, fog forms. Advection fog needs a push of air, usually 10 knots or more wind speed, to propel enough moisture towards land to reach saturation. Watch for close temperature/dew-point spreads in coastal areas, and anticipate the possibility of fog in the late afternoon and early evening, when the sea breeze has blown enough moisture inland for condensation to occur.

Upslope fog. When air rises, it expands, and this expansion causes it to cool. If there's a lot of vapor in the air, it might cool to the saturation point. This is what causes clouds to obscure mountain peaks and passes in late morning through afternoon on an otherwise clear day. Planning an airplane rental in Hawaii? Expect upslope fog to block the ridge lines by about 10 a.m.; by mid-afternoon, rain is a daily occurrence along the peaks. It will clear out again by morning.

There's another common upslope condition that receives little notice from flight service; this situation happens in the Great Plains. Field elevations gently but consistently rise from approximately the Mississippi River to the Front Range of the Rocky

Mountains. If air blows uphill, out of the east, it tends to gradually cool with the increase in altitude; eventually, an upslope fog might condense. Beware of winds out of the east in Oklahoma, Kansas, and Nebraska; they can bring a fog that lasts for days.

Because nights are longer, providing more time for cooling to take place, fogs are more common in winter than any other time of the year. They can, however, form anytime the air is relatively cool and a source of moisture is present. Watch for wind flows from bodies of water, or clear mornings after a strong rain or snowfall, which makes moisture available for evaporation in the morning sun. Airports tend to be built in river valleys and other low-lying areas, so expect fog near the runways even when surrounding areas are clear; I've even seen the moisture from a well-irrigated golf course (a seemingly natural occurrence near airports) create fog over the airport on an otherwise sparkling day.

MIST AND HAZE

Mist and haze are two more limiters to flight vision. The difference between a mist and a fog is the size of the associated water droplets. The heavier, moister drops of a mist form a thinner, greyer layer than fog, but they can result in airframe and carburetor icing if the temperatures are right. Mist is really a very light rainfall.

Haze, on the other hand, is associated with atmospheric pollution. In common parlance we call this smog. Minute dust particles that form the pollutant attract condensed water; the distribution of these particles is not uniform, and the miragelike "optical haze" effect can distort vision. Haze is especially problematic when looking directly into the sun. Haze tends to form in inversions, and consequently it has a definite "top," from a few hundred to a few thousand feet above ground level.

PRECIPITATION AND BLOWING

Heavy rain and, especially, snow can severely limit visibility. Strong surface winds can pick up dry snow, dust, or sand and create a cloud that obstructs vision on a local or even statewide level. Watch out for reports of heavy rain or moderate or greater snow showers. Beware of very low dew points in sandy regions or after a snowfall; combined with wind, these can drop visibility to near zero with little warning.

SMOKE

Are you flying to an airport near a large forest or grassland? A large wildfire can put enough smoke into the air to shut down airports for days. Is your destination in a largely agricultural area? Many farmers burn off fields in the autumn and the spring to prepare them for the next season's planting. Flying into a region marked by volcanoes? Even seemingly dormant peaks occasionally push a plume of smoke skyward. If you're flying into any open or agricultural area, always carry enough extra fuel to see you safely to an alternate, even in clear weather. You never know when Mother Nature or an industrious farmer will cause visibilities to drop near your destination airport.

REDUCED VISIBILITY WEATHER PRODUCTS

Many weather products available routinely or for the asking point to reductions to visibility aloft and at thc surface. Any report of thunderstorms, of course, implies the possibility of precipitation-induced low visibilities. Other sources of information include:

Weather depiction chart

The weather depiction chart (Fig. 10-1) is a pictorial compilation of surface aviation weather reports. It highlights areas of VFR, marginal VFR, and IFR conditions as they existed at the time of observation. Individual station symbols offer the actual visibility at each reporting point, as well as the phenomenon (fog, rain, etc.) limiting visibility.

Low-level significant weather prognostic

The "prog chart" (Fig. 10-2) presents the same VFR, MVFR, and IFR information as the weather depiction chart, but with a difference: this is forecast, not actual, data, for up to 12 and 24 hours into the future. Use this as part of your initial go/no-go planning, but expect things to change closer to flight time.

Pilot reports, area forecasts, and terminal forecasts

The reports (Fig. 10-3) report existing (PIREPs) and forecast areas of poor visibility. Pilot reports often include cloud base and tops altitudes, as well as flight visibility. Area reports indicate areas of widespread low visibilities anticipated in the forecast period; terminal forecasts are required to report expected visibilities less than 7 miles near airports. If visibility is forecast at less than 7 miles, the terminal forecast must also state the reason (haze or smoke, for example) visibility will be low.

Constant pressure chart

Constant pressure charts (Fig. 10-4) are compiled from those upper air observations at 00Z and 12Z. Reports are issued at various air pressure values, which approximate altitudes: 850 millibars is near 5000 feet MSL, the 700 mb chart approximates 10,000 feet, and the 500 mb plot reports conditions near 18,000 feet. You can ask for the 300-mb chart (30,000 feet), or the 200-mb plot (around 39,000 feet), if you need upper-air data.

Each station symbol on this chart is centered on a circle. If that circle is filled in, the temperature/dew-point spread at that altitude is less than 5 degrees: clouds are possible. To the lower left of the symbol, the actual temperature/dew-point spread measured at that altitude is given. If this value is less than five, and/or the station symbol circle is filled in, you might encounter clouds at that altitude.

What altitude do we mean? To the upper right of the symbol is a three-number identifier; put a "0" to the right of the third digit and read the altitude at which the radiosonde reached that chart's pressure level. If the symbol is filled in, that's an altitude where you can expect clouds. Conversely, you can use the constant pressure chart to

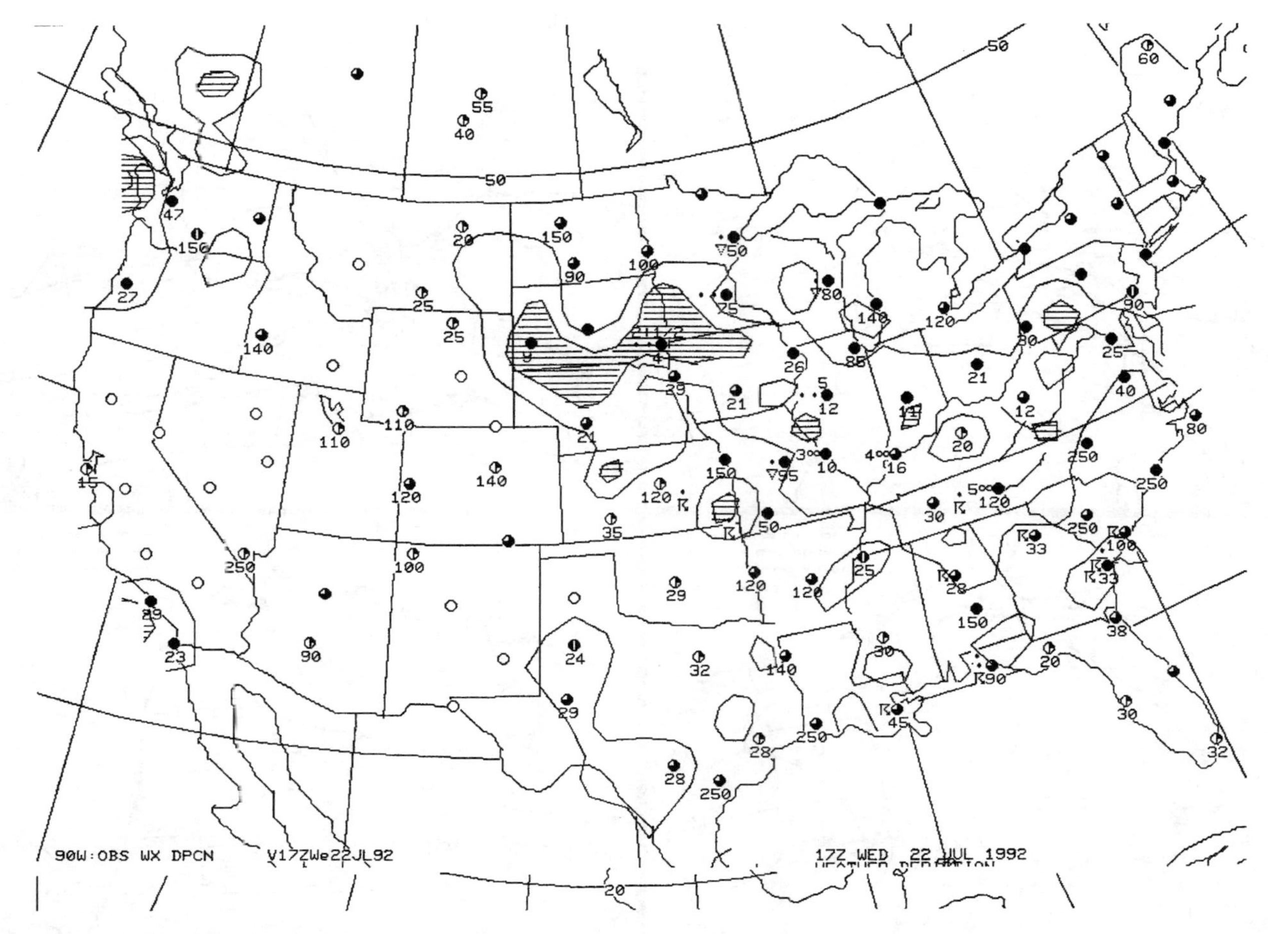

Fig. 10-1. *A weather depiction chart (Advisory Circular 0045B).*

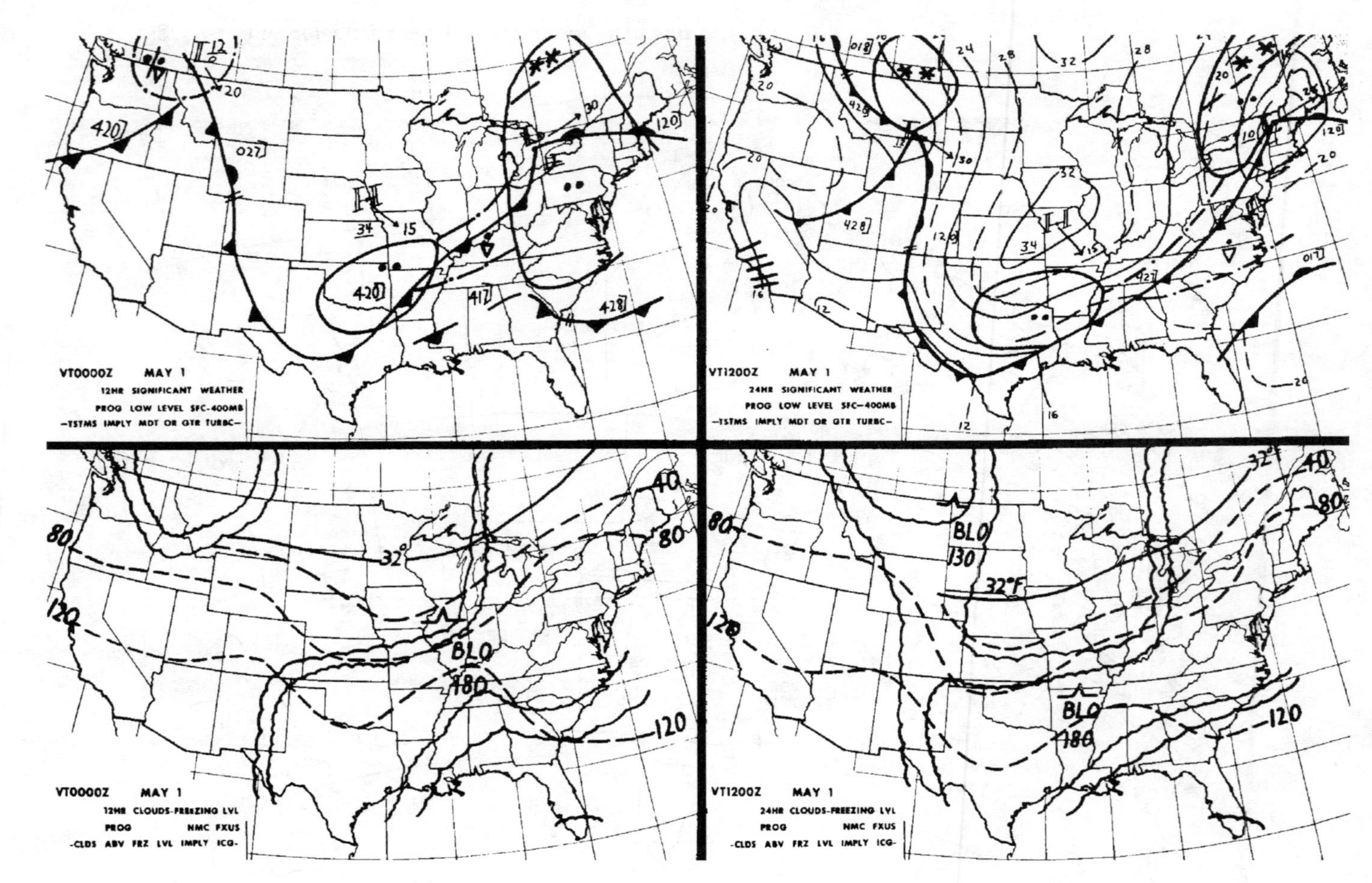

Fig. 10-2. *A low-level significant weather prognostic (Advisory Circular 0045B).*

```
MSP UA/OV FGT/TM 1600/FL020/TP OH58/SK 200 SCT/WX
FV7H/TA08/WV 19030/TB LGT OCNL MDT


CHIC FA 231045
MO
NRN HALF..CIG 10-20 OVC. VSBY 3-5S-. OCNL ZL-. TOPS 150. OTLK
MVFR CIG.
MSP FT AMD 1 181509 12 SCT C250 BKN 5H 1512G20 OCNL 2BS
```

Fig. 10-3. *A pilot report (top), an area forecast (center), and a terminal forecast (bottom), all indicating reduced visibility.*

ensure that you'll likely stay out of the clouds for a selected altitude. Don't have access to the hard copy? Ask your Flight Service briefer for the appropriate chart, and ask if "the station symbol is filled in" or, on their new "SuperFSS" computer screens, if the area on the chart is shaded in green. If the answer is no, expect clear air at that altitude; if it is yes, anticipate a layer of clouds.

Visibility reports

There are two methods by which visibilities are measured: human-based reports, and observations made by mechanical means. Human, or trained-observer, reports are based on known distances to prominent landmarks near an airport. Observers might know, for instance, that the water tower southeast of the airport is 1¾ miles away. If they can barely see the tower, they'll report 1¾ miles visibility; if they can't see it at all, they'll look for closer known landmarks to make the observation. Human-based reports are currently given in statute miles.

Sometimes visibilities will reduce to the point that the observer decides to measure visibility down a particular runway. This is especially important if the visibility is not uniform around an airport. The resulting runway visibility (RVV), reported in statute miles, is the distance down the stated runway that unfocused, unlighted objects can be clearly seen. This is again based on subjective, but trained, human judgment. If field visibilities are not uniform, the observer might issue a *sector visibility*, the visibility in a defined direction; a *prevailing visibility*, or that visibility typical to the airport at the time of observation, might also be given. This prevailing visibility is often also called the *tower visibility*, that noted from the tower cab at the time of observation. Coming into Charleston, West Virginia, in a snow shower last winter, I heard a visibility report like this:

> Tower visibility one and one-half miles, sector visibility ¾ miles to the northwest. Runway 22 runway visibility ¾ miles in moderate snow.

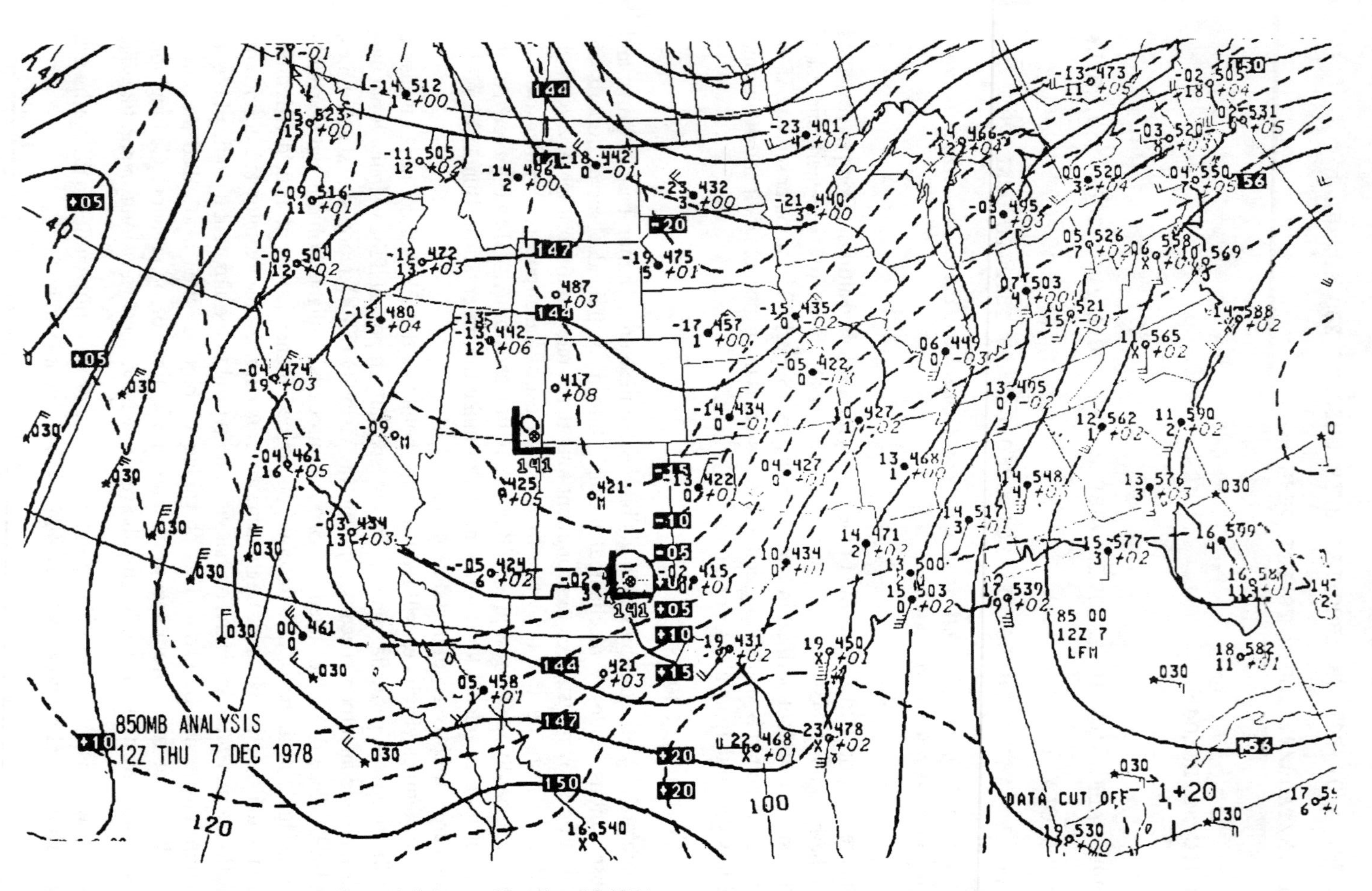

Fig. 10-4. *A constant pressure chart (Advisory Circular 0045B).*

This told me that things weren't too good for Yeager Airport, as the wind was out of the northwest; visibility was above ILS minimums but dropping rapidly; the airport closed due to visibilities as I taxied to the general aviation ramp.

Many larger airports have a mechanical means of determining visibility. Most common is a *backscatter* system that aims a beam of light across a small box exposed to the outside air; it measures the amount of light reflected back to a sensor and correlates that reflectivity into a distance measured in feet. This derives the *runway visual range*, or RVR, for an identified runway. You might hear a visibility report, then, like this:

Prevailing visibility ½ miles in fog. RVR runway 19 Right RVR 4000 feet.

The airport is at typical ILS minimums at all runways except 19 Right, and a little above minimums on the stated runway.

A little more about visibility reports: All nighttime visibility reports assume dark-adapted eyes. If the visibility is greater than six miles, it might be omitted from a weather report. If no mention of visibility is made, then, assume the field is "good VFR." If the visibility is six miles or less (it's required to be reported), then the source of the visibility limitation is also required to be given. If visibility at my destination is limited by fog, I'd like to know it; I can anticipate when the fog is likely to burn off, and if my arrival time is several hours in the future, I might proceed (taking along a healthy reserve of fuel). If the visibility is down because of smoke, and I hear a news report of an uncontrolled forest fire near my destination, I probably won't be able to land that day. Is the limitation caused by precipitation? I can ask the briefer when that rain or snow is likely to taper off or quit.

If visibilities are below legal or your personal minimums, of course you'll need to delay or reroute your flight. If conditions are close to minimums, or are changing rapidly, ask the briefer (or query DUAT or some other source) where the nearest improving conditions are. In other words, IFR pilots should never take off without knowing what direction will take them to MVFR or visual conditions, in order to plan an escape should an emergency arise. VFR pilots, on the other hand, ought to be aware of where IFR conditions exist anywhere near their route of flight, so as to be able to anticipate reduced visibility blowing towards their flight path and to know where not to turn if a diversion becomes necessary. By knowing what causes visibility to drop, getting the best weather information available, and asking healthy, yet skeptical questions, you can plan your flight to avoid the most deadly aviation weather hazard—reduced visibility.

11
Ice

ICE CAN ROB AN AIRPLANE AND ITS ENGINE OF VALUABLE EFFICIENCY AND capability. Currently, there's no way to positively detect the absence or presence of ice except for a pilot report; when icing conditions are present, then, everyone becomes a test pilot. Your job as a pilot is to anticipate the possibility of ice formation and to have a definite course of action in mind to make your escape should you actually encounter ice. There are two mutually exclusive types of ice that pilots need to plan for: carburetor icing and airframe icing.

CARBURETOR ICE

Many aircraft engines use carburetors to meter fuel into the induction system through a carburetor throat, or venturi (Fig. 11-1). If you remember back to your early days of ground school, you'll remember that air flowing through a venturi does three things: it speeds up, its pressure drops, and its temperature cools. The carburetor depends on the pressure drop to "suck" fuel into the induction air and to allow that fuel to atomize for combustion. That's why the carb has a venturi in the first place. The negative effect, of course, is that air in the venturi might be cooled as much as 40 or more degrees Fahrenheit; moisture in the carbureted air might freeze against the throttle plate, restricting throttle movement and air flow until the engine dies for lack of air (Fig. 11-2).

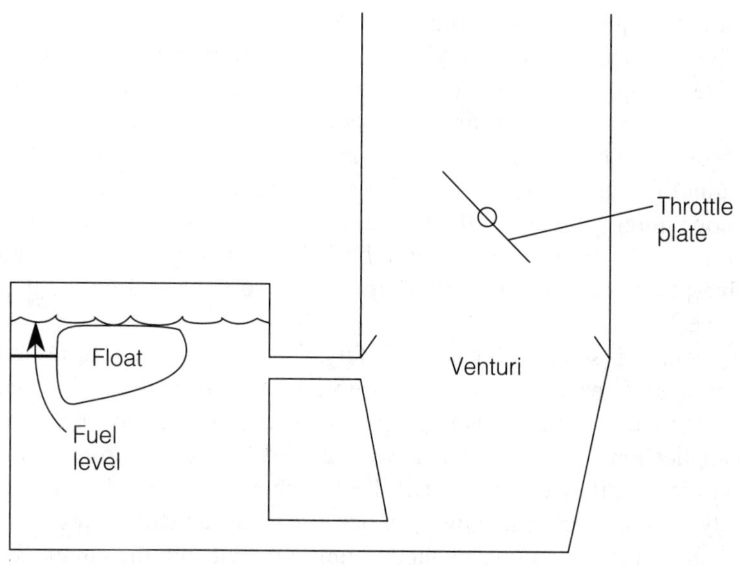

Fig. 11-1. *A simplified drawing of a float-type carburetor.*

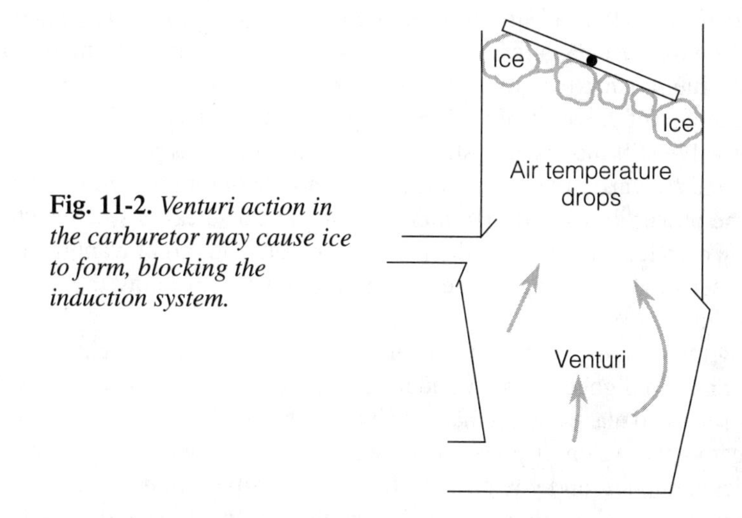

Fig. 11-2. *Venturi action in the carburetor may cause ice to form, blocking the induction system.*

How do you prevent carburetor ice? By running carb heat, you duct air heated by proximity to exhaust gases into the carburetor, hopefully preventing the formation of ice. This only works if you've selected carb heat prior to the accumulation of a significant amount of ice; if you wait too long, the carb heat might not be able to clear the induction system.

Whenever you're at low power settings (the engine temperatures will be lower, supporting the formation of carb ice) or anytime you're flying in extremely high humidities or any visible moisture (precipitation or clouds), you need to select partial or full carburetor heat (consistent with the *Pilot's Operating Handbook* for the airplane you're flying). The prime outside air temperatures for carb icing are from about 20 to 70 degrees Fahrenheit, so take OAT into account when deciding whether to use carb heat. Carburetor ice can form even outside of that temperature range if humidities are extremely high. I know a pilot who landed his Ercoupe in a milo field one muggy day when the temperature was 100 degrees; the engine failed because of ice in the carburetor. Anytime humidities are high, pay close attention to your engine power output and apply carb heat as soon as you notice the slightest drop in power.

As this control is so crucial to flying safety, you need to check carb heat as part of your before takeoff checklist. When I used to give a lot of biennial flight reviews in carbureted airplanes, I noticed that many pilots failed to completely check the effect of carb heat application. In the runup area, with around 1700 engine rpm set in the average Cessna product, they'd pull the carb heat knob and notice a drop in power, then immediately remove carb heat. Now, you don't want to run carb heat too long on the ground because in doing so you're ducting unfiltered air into the engine, but you do need to let it run long enough to see whether conditions are producing carb ice even at that point.

Apply carb heat and note the power drop. It goes down because the heated air you're diverting to the carburetor is less dense than what normally goes into the engine; less dense air means less oxygen is available for combustion, and the resulting, richer fuel/air mixture results in less power. So far, so good. But you should note the amount of power drop; it'll be roughly consistent from flight to flight if the carb heat valve still moves as it should, and you should keep carb heat selected for a few seconds to verify there's no increase in engine output before you remove carb heat. If the power drops and then increases while carb heat is still selected, you're picking up carb ice on the ramp, then melting it out. You might want to take off with carb heat selected, if your airplane's manual allows it, or delay the flight until the air is less humid.

Flying through visible moisture or in very humid air? You might want to test the carburetor heat in flight every now and then at cruise power, but otherwise just like you do on the runup, to make certain that power drops and stays at this lower level until you remove carb heat. Again, if power drops and then climbs while carb heat is selected, you're picking up ice, and you need to fly with the carburetor heat on.

In practice, I tend to approach this in a slightly different way. If I suspect possible carb icing conditions, I'll pull the carb heat knob partially out, advancing the throttle to regain the power I lost. Now I'm continually heating the carburetor throat. Every now and then I'll turn off the carburetor heat. I should get an immediate rise in power; if I don't, I know carb ice is forming, and I need to reapply full carb heat and find an altitude that'll give me warmer or drier air in the induction system. If I do get the expected power increase, I resume using partial carb heat anyway, just in case, and I test it again later.

AIRFRAME ICE

Regardless of the type of engine installed, all airplanes are susceptible to airframe ice. Many airplanes are certified for flight into "known icing;" we'll talk about that in a bit, but every pilot needs to have a plan of escape should he or she happen on airframe ice.

Unlike carburetor icing, airframe ice cannot accumulate unless you're flying through visible moisture such as a cloud, rain, drizzle or, in some cases, snow. Ice on the airframe has two effects on an airplane: increased weight and, by far the greatest threat, reduced aerodynamic efficiency.

I talked to a veteran Alaskan bush pilot one time who flew a Cherokee Six in the northern wilds. We were discussing airframe icing (on a different make of airplane), and he asked me what the maximum gross weight of that airplane was. "I like to fly a bit below maximum gross in the winter," he told me, "so I can handle the added weight of airframe ice." He went on to extol the superb weight-carrying capability of his usual mount, which, as he put it, "made it handle ice so well."

The added weight of ice on an airframe is the least of your worries should it begin to accumulate. Yes, the airplane might be capable of carrying the weight of the added ice, but long before maximum gross is exceeded, aerodynamics quite likely could have changed to the point that the airplane is totally unpredictable.

What happens when ice accumulates? Ice disrupts the normal air flow across airfoils. This is just like applying spoilers on a sailplane; lift is reduced, and the plane might not be able to hold altitude with the power available. Unlike the sailplane pilot, however, the pilot with airframe ice might not be able to remove its spoiling effect; couple this with the uneven distribution of ice, making lift development and stalling speed unpredictable, and the reduced power resulting from ice "spoiling" thrust on the propeller, and you can see a disaster in the making. I read a preliminary report of an accident in the Pacific northwest that cited a NASA study claiming that 1/16 of an inch of ice on a wing can reduce lift by 20 percent. Even a little ice, then, can be dangerous.

So airframe ice primarily reduces power and destroys lift. What else might it do? It coats propellers, reducing thrust just when reduced lifting efficiency requires an increase in power. Ice can spread rearward across airfoils and limit the movement of control surfaces. Airplanes with aileron and elevator gap seals seem to be prone to ice collection at those points, if the seals leak; if you're flying an airplane so equipped, check control movement frequently when flying in icing conditions. Airplanes with balanced rudder and elevator horns tend to pick up ice on those balances if the surface is moved from the streamlined position; this ice can lock the controls in place. If your airplane has overhanging control balances, you need to try to limit control deflections while you're accumulating ice. Although I have no scientific evidence to back this up, it stands to reason that vortex generators, which modify air flow by creating venturi action on the crest of the wing and other surfaces, will ice up quickly because of the drop in temperature that venturi action creates; this would form a ridge of ice exactly where it would do the most harm, in the wind flow on the top of the wing where even deicing systems couldn't remove it. I've asked several light-twin pilots who've installed vortex generators on their airplanes, and a few have reported ice formation here before any other spot on the airframe.

Finally, ice can limit the movement of secondary surfaces, such as flaps and trim tabs, and possibly freeze retractable gear in an up or a down position. If your gear is down when you accumulate ice, leave it down until you land or until you can be assured the ice has melted completely away. If you notice that trim tabs are stuck because of ice, don't try to break the ice free by forcing the trim actuator; you'll probably break some other component of the trim before you could dislodge the ice. You'll just have to do without the trim and the autopilot until you melt off the ice.

ICE FORMATION

When, then, is structural ice most likely to form? In order for ice to accumulate, the associated "visible moisture" has to be in liquid form. If the moisture in the air is so cold that it exists as suspended ice crystals, such as in high clouds or a dry, powdery snow, it will simply bounce off the airframe with no accumulation.

Airframe ice, then, is limited to a range of outside air temperatures, when the water is cold enough to freeze on contact with an airplane, but not so cold that it is completely frozen solid before contact. Most texts state that the prime temperature range for ice formation is from around +5 to –15 degrees Celsius. Below –15 degrees Celsius, water in most cases has completely crystallized. Why might you pick up structural ice if the temperature was as much as five degrees above freezing? It's possible that your airplane itself is very cold, from long periods cruising at cold outside air temperatures. You might have flown in clear air above the clouds but at below-freezing temperatures for hours; when you descend into moist air, your airplane might be cold enough to freeze water slightly above freezing before your passage. Add to this the slight cooling of air as it passes around the venturi represented by your wings, and you'll see why airframe ice is a hazard even at temperatures slightly above freezing.

Assuming a standard temperature lapse rate of 2 degrees Celsius per 1000 feet, you can plot the altitude range where ice is likeliest to form: from around 2500 feet below the freezing level to as much as 7500 feet above the freezing level. When planning a flight, then, select an altitude at least 3000 feet below the freezing level or 8000 feet above it, if possible, to avoid the likelihood of ice. Of course, if there's no visible moisture, there's no hazard in flying within the prime freezing range.

An exception to this temperature range is if the clouds are cumulus types. The rapidly rising air columns that form cumulus clouds can support larger droplets of water than stratus types and lift them so rapidly that they don't have time to freeze completely. This so-called "supercooled" water can exist in liquid form at altitudes where the outside air temperature is as low as 40 below, Celsius. If the air is unstable and cumulus clouds are likely to form, avoid flying through cloud tops above the freezing level, if possible, to avoid the hazard of ice (Fig. 11-3).

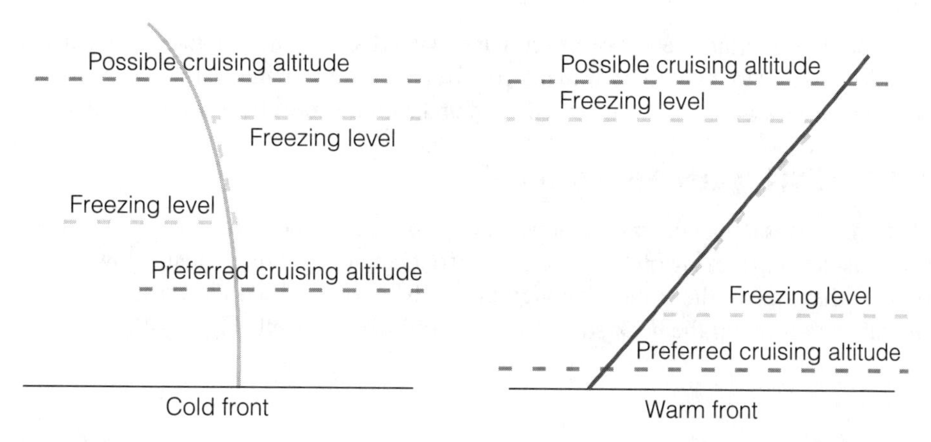

Fig. 11-3. *Plan to penetrate a front at an altitude away from the prime freezing range.*

TYPES OF AIRFRAME ICE

There are two types of airframe ice: rime ice and clear ice. Each has its own properties and can be anticipated in specific weather patterns. We've already touched on these patterns of ice development in the chapter on cloud types, but a review here is in order.

Rime ice is a granular, small-diameter type of ice usually encountered in stratus-type cloud formations. Because stratus clouds have little lifting action, they can't support large droplets of water; small droplets, when freezing, form small ice crystals that look like the frost on the inside of your kitchen freezer. Because stratus clouds tend to have less moisture than cumulus types, the rate of rime ice accumulation is usually low. Another nice attribute of rime ice is that it tends to come off quickly as soon as you reach clear air, often *sublimating* (going directly from solid to vapor state) at cruise speeds even when temperatures are still below freezing.

Anticipate rime ice anytime you expect stratus clouds: beneath troughs, in mist or fog, and along and ahead of warm fronts. Despite the low rate of accumulation, you might be exposed to rime icing conditions for a significant distance, so change plans to make an escape as soon as you notice an ice buildup.

Clear ice is formed by larger water droplets and so is found primarily in cumulus cloud formations. When these large droplets of ice impact a wing, they spread out across the flying surface and then freeze. Clear ice looks like an ice cube from your freezer. Clear ice, then, can limit the movement of control surfaces and freeze landing gear up or down, and it is difficult to remove, even in warm air. Since there is usually a good deal of moisture present in a cumulus cloud, the rate of clear ice accumulation is usually quite rapid. An escape is required at the first sign of clear ice buildup. Expect clear ice whenever you expect cumulus clouds—along cold fronts, near thunderstorms (even in the summertime), and in any convective activity.

You'll often hear a report of mixed icing. Mixed icing is a combination of rime and clear ice, usually found in slow-moving cold fronts or fast-moving warm fronts, the same weather scenarios that give you embedded thunderstorms in warmer times of the year.

FREEZING RAIN AND SLEET

Freezing rain is a phenomenon that occurs when a frontal boundary includes temperatures just above freezing on one side of the front, and temperatures just below freezing on the other side of the front. Although normally associated with a warm front, freezing rain can form on the back side of a slow cold front as well (Fig. 11-4).

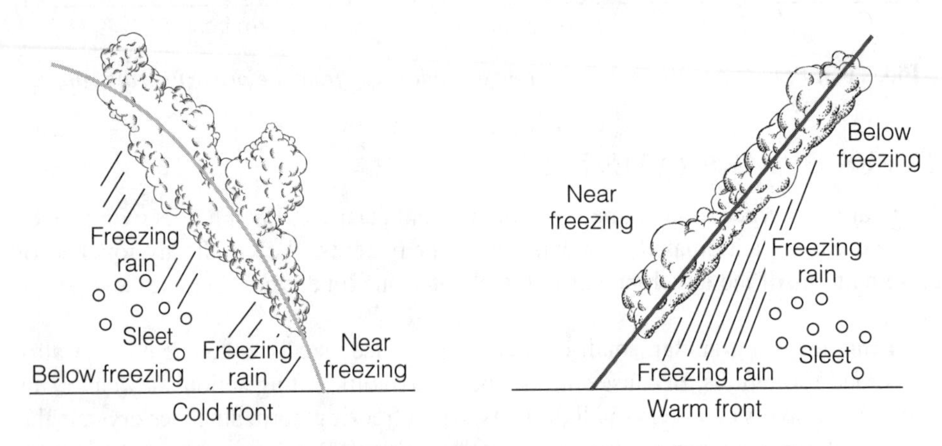

Fig. 11-4. *Freezing rain and sleet, or ice pellets, in a slow cold front (left), or a warm front (right).*

Freezing rain begins as liquid droplets, condensing and falling from the warm sector of the front. As the drop falls through freezing air in the cold sector, it also cools; when it contacts a cold surface such as the ground or an airplane in flight, the droplet breaks, spreads out, and forms clear ice. The rate of ice accumulation in freezing rain is so rapid that at least one airframe manufacturer advises that none of its products, even those certified for known icing, are designed to fly in freezing rain.

Freezing rain happens only if the drop has a short distance to fall in the cold sector, usually only a few hundred feet. If the droplet is exposed to freezing temperatures for a longer time, i.e., it falls through a greater distance in the cold, the drop itself freezes into what the National Weather Service calls *ice pellets*, what is commonly called *sleet*. If you hear warning of ice pellets or sleet along your route of flight, this assures you that freezing rain is present at some higher altitude. Overfly the area well above the freezing level, if you must go that route; better yet, delay your trip or divert to a different destination to avoid this extremely serious hazard.

WEATHER PRODUCTS THAT WARN OF ICE

By nature, thunderstorms can create extensive icing, even in summer. Carburetor icing can form anytime there is a lot of moisture in the air, so beware of high dew points or relative humidities and watch for the indications before and in flight. There are several charts and products, however, that warn of structural ice.

AIRMETs and SIGMETs

These reports, which we've discussed before, might also indicate the possibility of ice. AIRMETs are issued for reports of moderate icing, while SIGMETs are put out for areas of severe ice formation outside of thunderstorms (Fig. 11-5).

```
MIAZ WA 231445

AIRMET ZULU FOR ICG AND FRZLVL VALID UNTIL 232300

AIRMET ICG..NC SC GA

FROM TRI TO ECG TO 60SE ECG TO 50SW ABY TO CHA TO TRI

OCNL LGT-MDT RIME/MXD ICICIP BTWN 110 AND 240. CONDS CONTG

BYD 20Z.
```

Fig. 11-5. *An AIRMET calling for icing conditions.*

Pilot reports, area forecasts, and terminal forecasts

Pilot reports are the only means of verifying or refuting the presence of ice. Remember, it's not "known ice" until a PIREP is issued, and if you're the first to encounter ice, filing a pilot report is not an admission of a regulations violation. Report icing; not only is it required by regulation, it heightens safety for everyone.

The area forecast will include any warnings of airframe ice, as well as a statement of the freezing level. Watch for a "Z," indicating "freezing" on a terminal forecast's visibility section, as in "ZL" (freezing drizzle) or "ZR" (freezing rain). Either of those forecasts usually constitute a no-go item (Fig. 11-6).

Observed freezing-level chart

The observed freezing-level chart is another product of the 00Z and 12Z upper air observations. I like the freezing level chart for two reasons: it's actual, not forecast data, and it provides the actual freezing level in hundreds of feet above individual reporting stations. It's much more usable data, I think, than a forecast of a freezing level "8000 sloping to 10,000" (Fig. 11-7). Ask for the observed freezing level along your route of flight if you think ice might be a factor.

```
JBR UA/OVJBR240005/TM 1758/FLUNKN/TP PA34/LT RIME IN CLIMB

CHIC FA 231045

MO

NRN HALF..CIG 10-20 OVC. VSBY 3-5S-F. OCNL ZL-. OTLK MVFR

CIG.

JLN FT 231818 C25 OVC 3F 3017 OCNL C6 OVC 2ZL-F.
```

Fig. 11-6. *A pilot report (top), an area forecast (center), and a terminal forecast (bottom), all warning of icing conditions.*

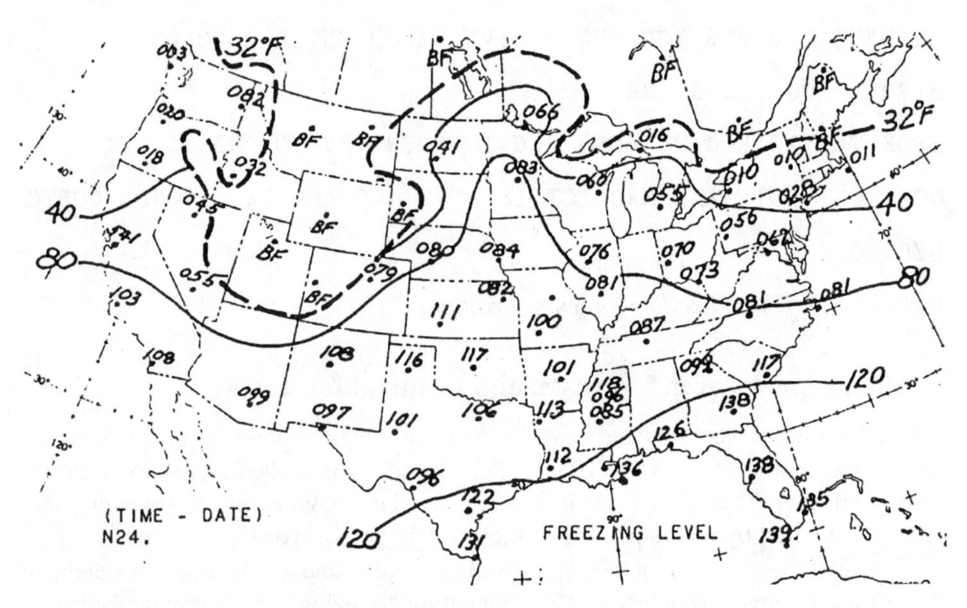

Fig. 11-7. *An observed freezing level chart (Advisory Circular 0045B).*

Low-level significant weather prognostic

Contrast the observed freezing-level data with what most briefings provide (the small lines on the upper portion of the "prog chart") and you'll see why I like the observed freezing-level chart better (Fig. 11-8).

Winds and temperatures aloft chart

You can, of course, get a pretty good idea of the approximate forecast freezing level from the winds and temperatures-aloft printout (Fig. 11-9). If you're getting your in-

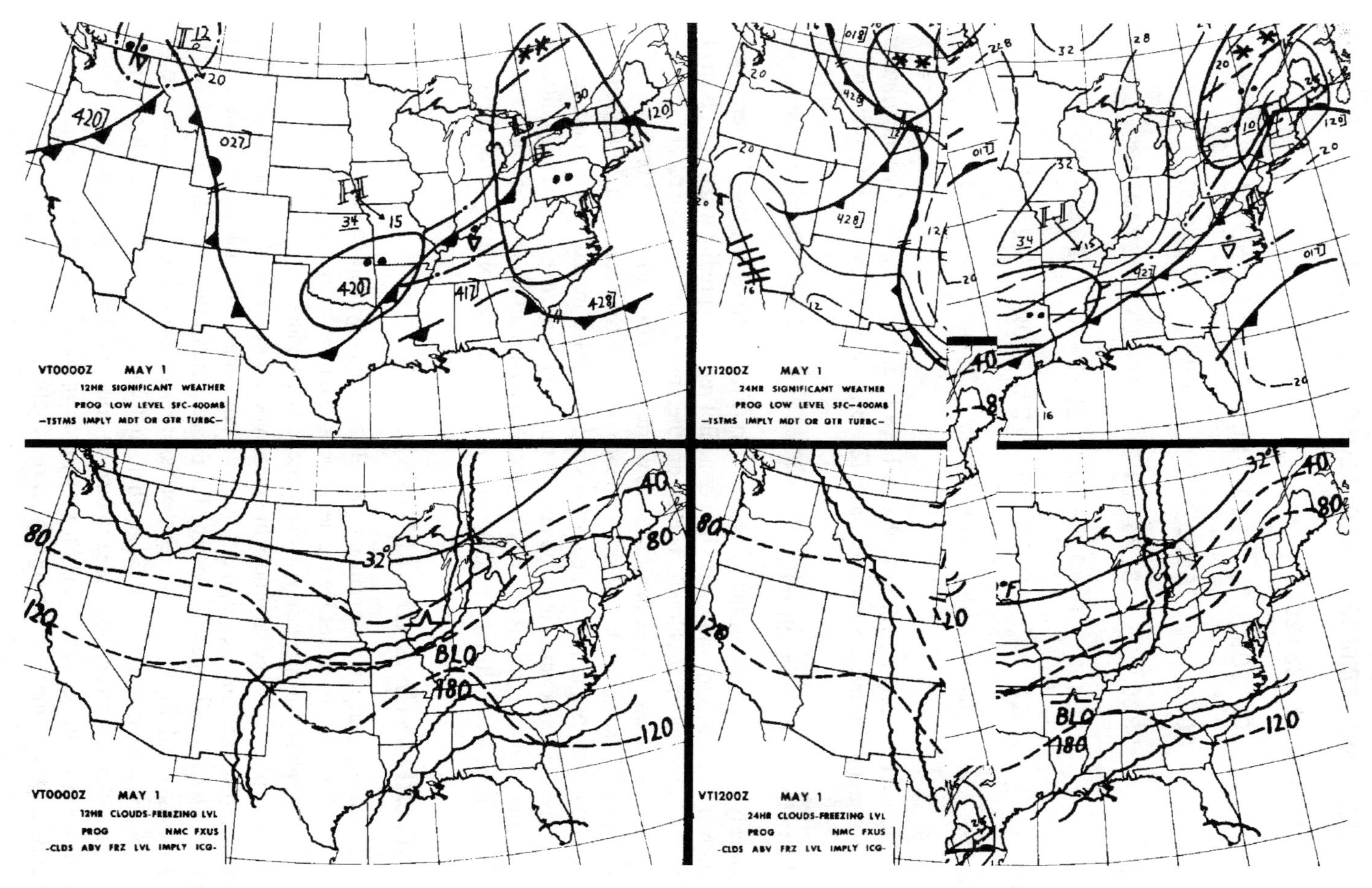

Fig. 11-8. *A low-level significant weather prognostic (Advisory Circular 0045B).*

FD WINDS AND TEMPERATURES ALOFT FORECAST

DATA BASED ON 231200Z

VALID 240000Z FOR USE 2100-0600Z. TEMPS NEG ABV 24000

FT	3000	6000	9000	12000
ICT	3320	3334-10	3340-13	3240-18
FSM	3126	3035-10	2943-10	2843-10
BHM	2516	2320+05	2235+02	2250-03

Fig. 11-9. *A winds and temperatures aloft forecast.*

formation over the phone, you'll probably have to ask specifically for temperatures aloft, even in the winter; in my experience, few briefers volunteer the information.

Constant pressure chart

We already looked at the constant pressure chart (Fig. 11-10) in the last chapter, where we learned that it was the only way short of a pilot report to predict whether a given altitude would put you in the clouds. There's another, highly informative part of the constant pressure station symbol that makes it worth your effort to ask the briefer for the information. To the upper left of the symbol you'll find the temperature Celsius reported at that chart's altitude. If the station symbol is filled in, indicating clouds at that altitude, and the temperature is in the normal range for icing, expect ice at that altitude. The constant pressure chart is the only product, except for an actual pilot report, that indicates airframe ice at altitude. Too bad we don't have a chart for every 2000 feet or so!

ICE AVOIDANCE AND REMOVAL

You're planning a trip in a Bonanza, on a cold autumn day, and you suspect ice in the clouds. So far, pilot reports of ice are negative, so you feel okay in starting out, but you want to have a plan just in case you encounter airframe ice.

Your airplane has one, and possibly two, types of equipment to deal with ice formation: anti-ice and deice equipment. Anti-ice equipment, like a windshield defroster, carburetor heat, or pitot heat, is designed to prevent the formation of light ice. It might not have the capability to remove ice that has already formed. If temperatures are going to be even close to the icing range, then, and you expect to fly through precipitation or a cloud, it's a very good idea to turn on your anti-icing equipment just prior to takeoff (consistent, again, with the *Pilot's Operating Handbook* restrictions), and leave the anti-icing equipment on through the entire flight.

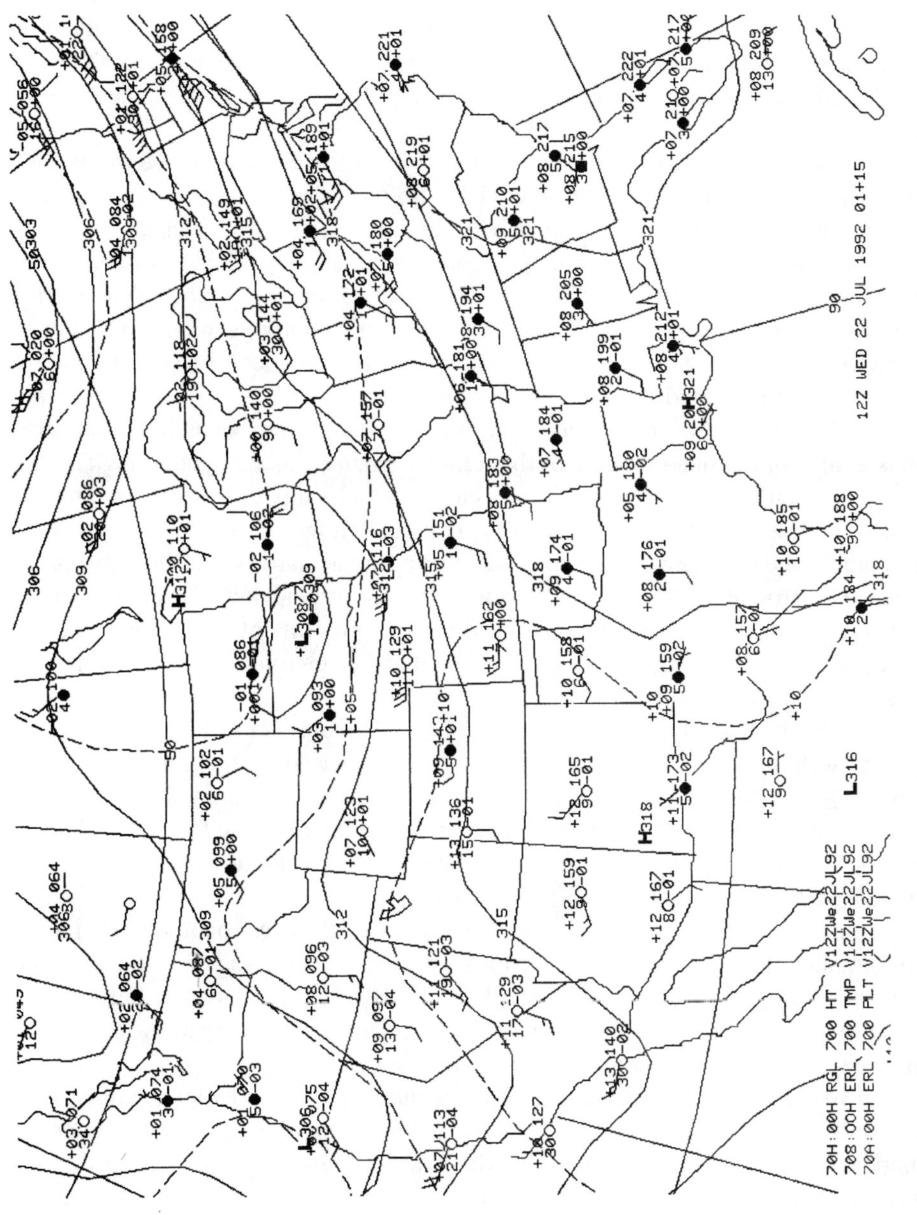

Fig. 11-10. *A constant pressure chart (Advisory Circular 0045B)*

You might have some deicing equipment on board as well. Typical deicing equipment includes inflatable leading-edge boots, "hot" propellers, and certain types of alcohol dispensers. These types of equipment are designed to remove ice accumulating at up to moderate rates; you don't need to turn them on until ice actually begins to form. If your propeller isn't ice protected, occasionally varying the rpm a hundred or two one way or another will help keep the propeller free of ice. See your *Pilot's Operating Handbook* and the appropriate handbook supplements for details about how to use your ice protection equipment.

You're en route now, immersed in a layer of stratus, when you notice ice has begun to form on the leading edges of the wings. So far it's just a trace of rime ice, but you don't know whether it will continue to build or not. You start to evaluate your options. You could, of course, continue on, hoping that the ice buildup will subside and that what you have already picked up won't adversely affect the landing. You remember, however, reading reports of pilots flying in similar conditions, with similar hopes and denial of the hazard, until it was too late. So you wisely decide to make your escape. Should you climb, descend, turn left, right or full about?

There's no set answer to that question. What you need to make an informed decision is a three-dimensional picture of the weather around you. Are you flying along a front, where you'll be exposed to the icing hazard for a long period of time, or are you crossing the front perpendicularly, soon to "pop out" on the other side? Are you near the frontal boundary, where temperatures above you are actually warmer, or are you far enough away from it that "up" is colder, not a good option, and down is your only hope of heating? Where are the cloud bases, and are there reports of tops? These are the things you need to know before you take off, and update en route, to deal with the potential hazard of ice.

Consider the pilot flying along a cold front, in the cold sector (Fig. 11-11). If she begins to accumulate ice, she has a couple of options to escape it. She can turn towards the front if she's not too far from the frontal boundary, and be assured of warmer air at the same altitude in a short distance, or she can turn away from the front if she knows the cloud bank ends nearby. She can climb into the warm sector as well, as long as the warm air isn't so far up that she'll pick up too much ice to maintain a climb, or she can descend, hoping to reach warmer or clear air at a lower altitude. She needs two pieces of information, then, to make her decision: where the front is located, and what the temperatures are at the surface and aloft. She could ask for the information before takeoff and remove a lot of the stress of an inadvertent ice encounter, by knowing beforehand what her best hope of escaping the ice will be.

You're in trouble in your Bonanza; ice continues to build, and your attempts at escape so far aren't working. With ATC assistance you begin the approach to a nearby airport. What technique should you use? Many nonice certified airplanes, like the Bonanza, have essentially identical counterparts like the Baron that are certified for known icing. Those airplanes will have in their handbook a minimum ice penetration speed. In the case of the Beech Baron, it's 130 knots. This speed has nothing to do with the reduced lifting capability and increased stall speed of the ice-laden wing; you have

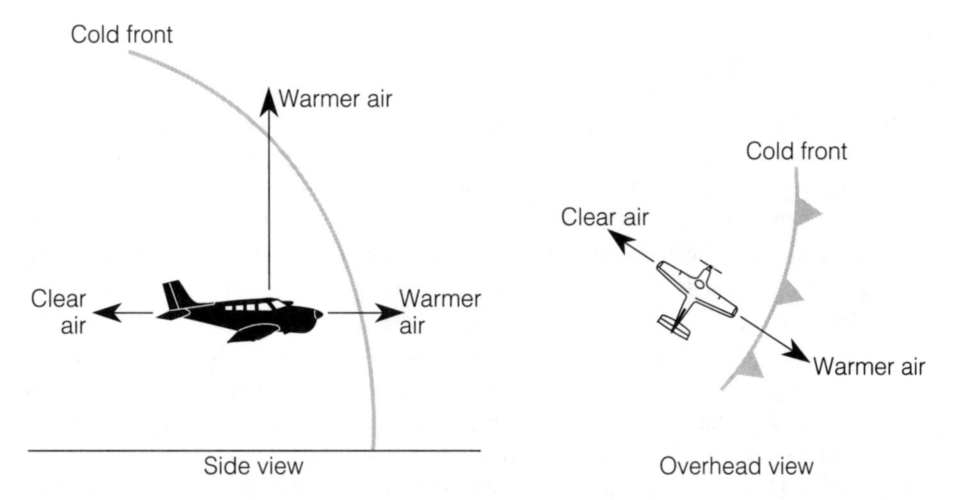

Fig. 11-11. *Evaluate your options before entering areas of suspected ice.*

no way of knowing what these values will be. Instead, this speed is one below which the airplane is at a high enough angle of attack that significant ice collects on the underside of the fuselage and wings. Flying faster than this speed means that less airframe ice will build. If your airplane is certified for ice, or is similar to a model that is, know its ice penetration speed and fly at least that fast if you encounter ice.

Should you use flaps? With ice on the airframe, lift is unpredictable, so at first you might think that flaps are a good idea. Using flaps, however, tends to increase the angle of attack of the airplane, creating an increased ice accumulation. Additionally, there is ongoing research (primarily related to commuter airliners, but applicable for all aircraft) that shows that the use of flaps, especially full flaps, removes some of the download created by the airplane's horizontal stabilizer, leading to an uncontrollable pitch nose downward. Within a few months prior to this writing at least two commuter airplanes have dived almost vertically into the ground in icing conditions, at about the point where full flaps are normally selected. It's a good idea, then, to avoid the use of flaps with any structural ice unless your *Pilot's Operating Handbook* specifically says otherwise.

Preparing for approach, reduce power gradually, an inch of manifold pressure or a hundred rpm at a time, to find an airspeed where you begin to notice a buffet or vibration. Then add power to increase airspeed at least 10 percent to give yourself a margin above stall. If you can practice this at altitude in clear air, all the better. Plan on holding at least that speed until the wheels hit the pavement; you're going to fly a power-on approach, and consequently use a lot of runway. If possible, pick an airport with a control tower, for this will usually ensure you of a long runway and crash-rescue folks just in case you need them. If you've got ice on the airplane, expect ice on the runway as well, so go easy on the brakes. It's better to roll off the far end of the runway at 20 knots than to stall and spin from too low an airspeed on short final.

KNOWN ICING

Pilot reports of ice in clouds, crucial to making an informed go/no-go decision, aren't common. A big part of this, in my opinion, is a lack of understanding about the regulations concerning flight into known icing, and the consequential fear among pilots that to issue a pilot report is to admit to breaking a regulation.

Icing is "known" when a pilot report exists for that general area and altitude. Pilot reports typically are considered to be valid for an hour, so if a single PIREP more than an hour before your arrival in an area calls for icing, you can launch into it in an non-ice certified airplane if you plan intelligently. If there's a trend of ice reporting, however, you probably should avoid the area even if the last report is more than an hour old.

If your airplane is not certified for known ice, and you encounter previously reported ice, you are required to issue a pilot report. Have you violated a regulation? No. It wasn't known ice until you reported it. Now it is known ice, so you have to escape it; you can't go back into that area, and the next guy can't take off into it, but you have not violated any regulations. Issue those pilot reports! Wouldn't you want to know before you took off? For that matter, get in the habit of issuing frequent pilot reports if you don't encounter ice. Knowing that ice doesn't exist in a cloud is a great worry-reducer, but knowing beforehand depends on everybody issuing reports.

Do you fly an airplane certified for known icing? Ice certification means that the airplane is capable of removing from protected areas ice accumulating no more than at moderate rates. We'll define those rates in a moment. Ice will still build on unprotected areas on an ice-certified airplane, so you'll still be in danger if you stay in even light icing for a prolonged time; basically, what ice certification does is to allow you to legally go into areas where ice is probably present and have the means of getting yourself out of trouble (i.e., escaping) should ice begin to form. Ice certification is not meant to lull you into remaining in icing conditions.

ICING REPORTS

We've touched on some terms used to identify the rate at which ice accumulates. Here are the actual FAA/Flight Service definitions of icing intensity:

Trace. A *trace of ice* indicates perceptible accumulation. The icing is hazardous to nonice-certified airplanes if they remain for more than one hour.

Light. *Light icing* means that occasional use of deicing equipment is required. Because of ice accumulation on unprotected areas, exposure to light icing for more than an hour can be hazardous even to ice-certified airplanes.

Moderate. *Moderate icing* means that continuous operation of deicing equipment is required to clear protected areas. In the case of deice boots, this means at least ½ to ¾ inch of ice on the leading edges, rebuilding almost immediately after cycling the boots. So exposure for even short periods of time is dangerous.

Severe. Continuous use of deicing equipment can't prevent buildup in *severe icing*, so immediate escape is necessary.

Remember, pilot reports are the only method in use today to detect the absence or presence of airframe ice. Get in the habit of filing pilot reports confirming or denying icing conditions you encounter en route, and have a plan in mind ahead of time to carry out your escape should you come across airframe ice.

FROST

Frost forms on exposed surfaces in cold, clear weather when the temperature/dew-point spread is close but not enough moisture is present to create a cloud or fog. Frost, like airframe ice, disrupts air flow across a wing; because frost forms as irregular, jagged crystals on the wing surface, it creates even a greater airflow disturbance than ice accumulated in flight.

The regulations concerning frost removal from the airplane are rather cryptic. In fact, Part 91 (noncommercial) flight regulations don't even address the issue. In commercial operations, the regulations talk of completely removing the frost or "polishing it smooth," whatever that means. You'd have to be certain that your "smooth" frost layer exactly matched the design contours of the wing to give predictable performance. I once patronized an FBO that suggested pouring buckets of water over my Skyhawk's frost-covered wings, to "fill in the frost" and make the wing airworthy; I won't give that FBO business any more. The better choice, by far, is to eliminate the problem by completely removing the frost.

There are numerous commercially available frost removers that you can spray over the airplane to remove frost, and they generally do an excellent job. Don't dawdle after deicing, however, because the frost that melted might refreeze into clear ice before takeoff. The same goes if you tow the airplane into a heated hangar to melt the frost. Be sure that the airplane is completely dry before towing it back out into subzero temperatures, or the icing hazard might be magnified. Be careful not to spray these frost removers on your windshield or windows, as they tend to craze the Plexiglass.

Finally, you could spring for a Glycol treatment like the "big boys" do, but be wary of water refreezing in extremely cold temperatures (despite the popular notion, this has led to a few accidents), and, again, avoid letting this substance touch your windows.

One last note about ground-based ice or frost removal: be extremely careful to verify full freedom of movement of all control surfaces and trim tabs before takeoff. I know of a King Air that crashed after deicing in a heated hangar; the industrious line crew dried every bit of the airframe except the hard-to-reach T-tail before towing it into the cold, and ice blocking the elevator movement led to an accident.

Airframe ice is perhaps the most feared and least understood aviation weather hazard because of its unpredictability and misconceptions about the definition and legality of "known ice." Asking the right questions when obtaining weather information, however, will allow you to anticipate the probable type and location of ice formation, as well as to construct a three-dimensional weather picture to avoid ice accumulation and to escape it if your attempt at avoidance fails.

WEATHER PRODUCTS THAT IDENTIFY HAZARDS

As a review, Table 11-1 shows a list of the aviation weather products I've mentioned in chapters 8 through 11; the table also indicates the primary aviation weather hazards information you can glean from each product. Of those products listed in the table, a standard weather briefing will typically include:

- Weather depiction chart.
- Surface aviation reports.
- Radar summary chart.
- Pilot reports.
- Area forecast.
- Terminal forecast.
- Significant weather prognostic chart.
- Winds (but not temperatures) aloft.
- AIRMETs, SIGMETs, and convective SIGMETs.
- Severe weather watch bulletin.

Ask for the appropriate other charts or reports if any of the above indicates the possibility of a hazard. For instance, if you suspect thunderstorms based on the area and other forecasts, knowing the stability index will help you make a better-informed go/no-go decision. Ask questions!

Table 11-1. Weather products that identify hazards

Product	T-storms	Turbulence	Visibility	Ice
Observations:				
Weather Depiction Chart			X	
Surface Aviation Reports	X	X	X	X
Radar Summary Chart	X	X	X	
Freezing Level Chart				X
Stability Index	X	X		
Constant Pressure Chart			X	X
Pilot Reports	X	X	X	X
Forecasts:				
Area Forecast	X	X	X	X
Terminal Forecast	X	X	X	X
Significant Weather Prog	X	X	X	X
Winds/Temps Aloft		X		X
Tropopause Ht/Vertical WS		X		
AIRMET/SIGMET	X	X	X	X
Convective Outlook	X	X		
Severe Wx Watch	X			

12 Weather-related aircraft accidents

WHY DO WE STUDY WEATHER? BESIDES WANTING TO KNOW WHAT CLOTHES TO pack, or whether we need to hangar our airplane the night before a trip, we watch the skies in order to be safe. Weather hazards, as it turns out, account for a hefty percentage of all general aviation accidents. We all know about the frequency of the "continued VFR flight into instrument conditions" sort of mishap, but that is only one weather-related accident scenario. This chapter is not meant to condemn any pilot or class of pilots, any airframe or other manufacturer, or any weather information outlet for actions or inactions that lead to accidents. Instead, it's written to educate those of us left behind in the hopes that we will benefit from others' experiences. Remember, all pilots are at risk, regardless of their ratings, experience, or the equipment on board their airplanes.

According to the Aircraft Owners and Pilots Association's Air Safety Foundation, 7.8 percent of all general aviation accidents in the years 1982–1988 were assigned "weather" as a causal or contributing factor by the National Transportation Safety Board. Of the single-engine, fixed-gear accidents whose causal factors were pilot-related ("pilot error"), 10.4 percent involved weather. Weather was a factor in 19.1 percent of all pilot error multiengine aircraft crashes and 19.7 percent of all single-engine retractable-gear airplane mishaps that were pilot-related. Single-engine retractable pilots had the greatest percentage of weather-related accidents, followed by twin-engine flyers. A predominant number of the accidents were attributed to initiating or continuing VFR flight into Instrument Meteorological Conditions by noninstrument-rated pilots, followed by instrument rated pilots attempting to continue VFR flight into IMC.

In 1991, the AOPA Air Safety Foundation published the *General Aviation Accident Analysis Book*, the first compilation of general-aviation specific accident statistics. Covering the years 1982 through 1988, this book reports the findings of the National Transportation Safety Board or designated agencies. The information is broken down into three groups (single-engine fixed gear, single-engine retractable, and multiengine airplanes), as well as the primary cause category (pilot error, mechanical, and other/unknown). Within these groupings, individual accidents are listed by phase of flight and primary and supporting causes, and the accidents are identified by specific aircraft type. The level of aircraft damage, and the number of fatal accidents and fatalities, are also listed.

Looking at these statistics, I found that a larger percentage of accidents could actually be attributed in part to weather hazards. The AOPA conclusions did not cover some cases where weather phenomena were a significant contributing factor, such as when a pilot lost control of the airplane on landing ("pilot error-loss of directional control" the AOPA's category) due to gusty or crosswind conditions. If you account for all accidents where weather was a contributing factor, you find the percentages listed in Table 12-1.

Table 12-1. General aviation weather-related aircraft accidents, 1982–1988

Airplane type	Percentage of total
Single-Engine Fixed-Gear	13.5%
Single-Engine Retractable-Gear	19.9%
Multiengine	16.7%
All Airplane Types	15.0%

Significantly, 32.3% of all weather-related accidents were fatal.

Pure statistics are meaningless unless put into some usable context. In order to help pilots recognize exactly what sort of weather-related situations have been the most hazardous historically, I've delved through the AOPA/NTSB data and grouped accidents by actual cause. By knowing these "cause categories," you might more easily recognize the risks associated with flying in and around weather hazards and avoid a repeat of accident history. What, then, are the most common weather-related accident causes?

ACCIDENT CAUSE CATEGORIES

Virtually all of the weather-related accidents were identified as "pilot error" by the NTSB. In other words, the primary cause of the accident was a pilot's improper action or failure to accomplish a proper action. A large number of these mishaps could have been prevented before the pilot ever left the ground, in the flight planning stage.

Occasionally the NTSB found that factors beyond the pilot's control were primarily responsible for a crash. Sometimes wreckage was simply never found, and no determination of the accident's cause could be made. With weather-related mishaps, these were classified in the "other/unknown" cause category. This situation was extremely rare. Table 12-2 shows the percentages.

Table 12-2. Nonpilot error weather-related accidents, 1982–1988

Airplane type	Percentage of total
Single-Engine Fixed Gear	1.7%
Single-Engine Retractable Gear	1.0%
Multiengine	2.1%
All Airplane Types	1.6%

Only 1% of all airplane fatal accidents were in the "other\unknown" category.

Ranked in order of the total number of accidents in the years 1982–1988, here are the weather-related causes of accidents. Percentages have been rounded to the nearest tenth and therefore might not total exactly 100 percent.

Turbulence/wind

Turbulence aloft, not associated with thunderstorm penetration, and gusty or crosswind conditions near the surface account for the greatest number of weather-related airplane accidents. In fact, 44.7 percent of all meteorological mishaps fall into this category. What does this mean?

If I had asked you, before you read this book, to tell me which of the four aviation weather hazards (thunderstorms, turbulence, reduced visibility, or ice) caused the most accidents, which would you have chosen? Most pilots I ask reply that reduced visibility is the greatest hazard, citing the familiar (and don't get me wrong, quite dangerous) "VFR into IMC" accident as the most common. What happens, then, is that pilots "set themselves up" for a turbulence-related accident, simply because they think that it's not nearly the hazard that it is. Adverse wind conditions lead to nearly half of all weather-related accidents. Now that you know that, do you think you'll pay just a little more attention when a Flight Service briefer, an ATIS broadcast, or a dancing wind sock warns of possible turbulence aloft, wind shear, or shifting winds near the ground?

That's the bad news. The good news is that this most common weather-related accident cause ranks second to last in the number of fatal accidents; only 18 of 1086 total turbulence/wind-related accidents were fatal. That's only 2 percent. Here's how these accidents took place.

Landing mishaps

As you'd probably expect, most of the turbulence and wind-related accidents happened close to the ground, where the pilot had little room to recover. In fact, 75.8 percent of the turbulence/wind crashes occurred during the landing phase of flight. Table 12-3 shows the specific scenarios.

Table 12-3.
Turbulence/wind landing mishaps

Accident cause	Accidents	Fatal accidents
Loss of Control	701	7
Hard Landing	87	0
Landed Short	35	0

Possibly predictably, this sort of accident was weighted abnormally heavily towards single-engine, fixed-gear airplanes, maybe the result of accidents among student and low-time pilots. All pilots should remember, though, that they are at risk in shifting winds, and they should delay or divert if conditions are beyond airplane and personal wind minimums. Consult your airplane's *Pilot's Operating Handbook* for crosswind computations and technique, and spend some time regularly with an instructor qualified in your particular model of airplane to maintain and improve your ability to handle turbulence, gusts, and crosswinds.

Takeoff mishaps

Similarly, 223 accidents, 20.5 percent of the turbulence and wind total, happened in the takeoff phase. The percentage of fatal accidents was higher, probably because the airplane is typically at a slow speed, with little margin above a stall, and close to the ground in this phase of flight. Table 12-4 lists accident causes. Again, the remedy is to know specific airplane procedures and techniques and to practice with a qualified instructor to master flying the particular type of airplane and to determine your own limitations.

Table 12-4. Turbulence/wind takeoff accidents

Accident cause	Accidents	Fatal accidents
Loss of Control	214	6
Improper Procedure for Conditions	9	1

Miscellaneous

The remaining 3.7 percent of turbulence and wind-related accidents are listed in Table 12-5.

Table 12-5. Turbulence/wind miscellaneous accidents

Accident cause	Accidents	Fatal accidents
Wind Shear (Location Unspecified)	26	0
Wind Shear (Missed Approach)	11	2
Loss of Control in CAT	2	1
Inflight Breakup: Excessive Speed on Approach in Turbulence at ATC Request	1	1

VFR in IMC (pilot rating unspecified)

Before you think, "Okay, here's the guy without a rating flying into the clouds," bear this in mind: this category is separate from those where lack of an instrument rating was cited as a factor. In other words, although the information I worked from lacked the detail to tell me exactly how many, the majority of these accidents happened to an instrument-rated pilot who flew VFR into IMC and then lost control or encountered terrain. Consider the crash of a single-engine, retractable gear airplane, reported in the May 1, 1994, issue of *Aviation Safety* magazine.

> The pilot called the Hawthorne AFSS (automated FSS) late in the morning and asked for the current weather conditions at Long Beach, California. He was told that Long Beach was reporting scattered clouds at 2000 feet, an overcast at 3300 feet and seven miles' visibility. He then filed an IFR flight plan for the short flight to Long Beach from Catalina Island.
>
> The pilot, 66, had a commercial license and an instrument rating. Details on his recent flying activity are not known, but he had reported 3300 total hours, including 100 hours during the previous six months, on his most recent medical examination form.
>
> He arrived at the Catalina Airport about two hours after his weather briefing (the accident report termed it a "self-limited" briefing). There was fog in the area; the sky was obscured and the ceiling was indefinite.
>
> While chatting with the airport manager, the pilot mentioned that although he had an IFR flight plan on file, he was considering departing VFR on Runway 4. The airport manager recalled that visibility to the northeast of the airport was about 1 mile.
>
> The pilot did elect to depart VFR, but he taxied (the airplane) to Runway 22, rather than Runway 4. The fog off the end of Runway 22 apparently was much thicker. The airport manager estimated that visibility to the southwest of the field was only about a quarter mile. The pilot announced on the un-

controlled airport's unicom frequency that he was ready for departure. The airport manager gave him the altimeter setting and advised that the winds were calm.

A few moments later, the airplane was in a shallow, right turn and in a slightly nose-low attitude when it crashed below the crest of a dirt road about a half mile northwest of the airport. The pilot and his passenger were killed.

The accident report noted that investigators found no evidence of any preimpact malfunction of the powerplant, airframe, or flight controls.

NTSB concluded that the probable cause of the accident was improper preflight planning. The pilot chose the wrong runway for departure, intentionally flew (VFR) into instrument conditions (without a clearance), and failed to select a proper altitude for terrain-clearance, the board said.

In all, 20.6 percent of all weather-related accidents fall into this category. It's also "number one" on the list of fatal accidents. Someone was killed in 401 of these mishaps, and 80 percent of all accidents of this type were fatal. What's the moral? Unless you plan to fly only in the best of weather, over short distances that remove a lot of the variables in weather forecasting, and without any business or family concerns to force a schedule, get an instrument rating, fly properly equipped airplanes, and train regularly to keep your "ticket" truly current. Then file IFR whenever the weather is marginal, and plan altitudes and routes (including escape paths) for the entire trip whether on an IFR or VFR flight. Table 12-6 shows the causes of this class of accident during the study period.

Table 12-6. VFR in IMC accidents: pilot rating not specified

Accident cause	Accidents	Fatal accidents
Continued Flight into Known IMC Weather	226	192
Entered IMC Flying Under Overcast in Rising Terrain	94	76
Initiated VFR Flight into Known IMC	75	64
Attempted VFR Descent/Approach in IMC	61	35
Attempted VFR Climb in IMC	44	34

Even a lot of instrument-rated pilots made bad preflight and in-flight decisions, forgot that safety is the primary goal of flight, and chose to begin or continue VFR flight without proper planning or when conditions were beyond their abilities.

VFR into IMC: (pilot not instrument-rated)

This is it, the category of weather-related accidents most pilots feel is most common. In these crashes, the lack of an instrument rating was specifically noted by the investigating agency. All 253 cases, 10.4 percent of the weather-related total, were the result of bad decision-making on the part of the pilot. In fact, the AOPA Air Safety Foundation states that in virtually all of these accidents, known IMC conditions were in existence at the takeoff airport or en route, and the pilot either failed to obtain a required weather briefing or chose to ignore the information received. That's unfortunate because 85 percent of these accidents were fatal. Table 12-7 gives the particulars.

Table 12-7. VFR into IMC: pilot not instrument rated

Accident cause	Accidents	Fatal accidents
Initiated Flight into IMC	131	120
Continued Flight into IMC	84	68
Attempted Approach in IMC	34	25
Attempted VFR Flight Above IMC	4	2

This is perhaps the most avoidable cause of airplane accidents, beginning or continuing visual flight into instrument conditions. Pilots regularly defy the logic of obtaining required weather briefings; those who do contact an official weather outlet might be jaded by the perceived "VFR flight not recommended" stance of some weather briefers. How can a pilot avoid this situation? Get a thorough briefing, and then take a critical look at actual conditions. It's okay to "go up for a look" as long as you have a definite escape route in mind, one that you know will allow you to "maintain VFR" and get safely back on the ground.

Finally, even if you don't intend to ever file IFR and launch into the murk, get an instrument rating, or at least a healthy dose of instrument instruction, to have "in your pocket" should conditions deteriorate too rapidly. It might well save your life and the lives of your passengers.

Improper IFR altitude or procedure

In the aviation industry, we call this sort of accident "controlled flight into terrain." What that means is that the airplane crashed into terrain or an obstruction under control, in level flight or in a normal flight attitude. This type of mishap is the result of the pilot flying an altitude or procedure that's inappropriate for his or her location in this ranking under Instrument Flight Rules.

There were 232 accidents of this sort in the years 1982 through 1988. Seventy-two percent of these accidents were fatal. Table 12-8 provides the specific situations.

Table 12-8. Improper IFR altitude or procedure

Accident cause	Accidents	Fatal accidents
Descent Below Minimums During Approach	114	69
Improper Approach/Transition Procedure	65	52
Improper Takeoff/Climb Procedure	31	27

Table 12-8. Continued

Accident cause	Accidents	Fatal accidents
Improper Missed Approach	21	19
Improper Enroute Altitude	1	1

This is a call for improved training and currency requirements for instrument-rated pilots, as well as a plea for individual pilots to realize that going "just a little bit" lower or further can indeed be fatal. Avoid repeating this sort of accident history by receiving regular, challenging instrument competency checks, preferably using approaches other than those with which you are most familiar. Also, consciously remember that your primary goal in flying is safety. Don't sacrifice safety to achieve convenience (arriving on time at the intended destination). The statistics show that often the price of "convenience" is too great.

Landing in marginal or IMC weather

Although only 4 percent of these crashes were fatal, 96 accidents occurred during the study period and were attributed to landings in marginal weather or after a successful instrument approach in IMC. Table 12-9 lists the causes.

Table 12-9. Landing in marginal or IMC weather

Accident cause	Accidents	Fatal accidents
Improper Off-Airport Site Selection: Precautionary Landing Due to Weather	65	0
Landed Short: Marginal Weather at Night	25	4
Landed Long After IFR Approach in IMC	8	0

This category is skewed by the large number of unsuccessful off-airport landings made because of threatening weather conditions. This suggests a situation that could nearly have resulted in VFR flight into IMC or IMC flight into known weather hazards. The case could be made that this sort of accident came about from a good decision ("get the airplane on the ground") made after a bad decision ("continue flight into weather hazards"). Of course, these accidents were really the result of improper preflight planning or a delay in making an in-flight decision until events became critical. I know the weather would have to be very bad for me to decide to put a perfectly good airplane down in somebody's field. The remedy? Get those required weather briefings, and use your knowledge of weather patterns and behavior to verify or refute the "real" conditions. Make a decision to divert to an alternate before conditions get so bad you elect to land off-airport.

Another large category of accidents happened in marginal (not instrument) weather conditions at night. Night landings involve different sensations and perspectives than those you experience in the daytime. Marginal weather creates additional distractions. Wet windshields might bend light rays, making runway lights appear to be closer or farther away than they actually are. Overcast skies or marginal visibilities create very dark conditions, especially at remote airfields; a rectangle of runway lights isolated in a "black hole" devoid of lights might appear to move, and depth perception is all but nonexistent. Be very wary of flying at night in less than good VFR weather, for all your daytime and IFR experience won't prepare you for a marginal night's peculiar conditions. Practice a lot of night takeoffs and landings on clear nights before you venture into the dark on a marginal evening.

Ice

Probably the least accurately predicted and the most often dismissed of aviation weather hazards, ice was responsible for 68 accidents in the years 1982–1988. Sixteen percent of the ice-related accidents were fatal. Table 12-10 shows the specific causes.

Table 12-10. Ice

Accident cause	Accidents	Fatal accidents
Stall on Landing	50	7
Failure to Use Carb Heat	14	3
Takeoff into Known Ice	4	1

These accidents were weighted disproportionately towards the single-engine retractable and especially multiengine airplanes. Perhaps increasing aircraft performance exposes pilots to icing conditions more frequently, or maybe it provides a false sense of confidence when a pilot begins to encounter ice. I remember that I was once scheduled to take a winter trip in a Bellanca SuperViking, only to have to switch at the last minute to a Cessna 172 because of airplane maintenance. One of my first thoughts was that I had better watch out for airframe ice. On reflection, I decided that maybe the switch "downward" in performance was healthy for me. I should have been just as concerned about ice in the Bellanca as I was in the Skyhawk.

Neither type is equipped or approved to handle any airframe ice. A common pilot comment concerning this hazard is that one type of airplane or another can "handle" a load of ice. The effects of airframe ice are extremely unpredictable. Ice accumulation is not uniformly distributed, nor does it conform to any rules of efficient aerodynamics. Pilot overconfidence about the ability of a particular design to fly with a load of ice is usually centered around that design's payload capability. The statistics, however, don't even mention hard landings or overstressed airframes, the result of a too-heavy aircraft. It's aerodynamics, expressed in the number of accidents related to an ice-laden stall, that bring down airplanes, and you can't predict the aerodynamics of an airplane coated with ice.

A few single-engine and numerous multiengine designs are certified to fly in "known icing" conditions. Pilots of those types should remember that ice certification simply allows greater flexibility in flight planning and a "safety valve" in escaping actual icing conditions; ice certification is not designed to allow continuous flight while accumulating ice. Accidents attributed to ice accumulation, as we'll see later, are much more common in multiengine airplanes than their single-engine counterparts, despite the prevalence of "icing certified" twin-engine designs.

The majority of these crashes occurred on landing, when the unpredictable aerodynamics of airframe ice caused an unexpected (and usually quite violent) stall when slowing for final approach. The aviation community is continuing to study the effects of ice on an airframe, and new information is constantly coming to light. For instance, it's a recently understood fact that ice accumulates on the tail surfaces of an airplane faster than on the wings, and that the use of flaps can cause an ice-contaminated tail to stall, resulting in an abrupt nose-down attitude. This is why the industry now recommends against the use of any flaps when any airframe ice is present, consistent with manufacturer's recommendations.

Should you "ice up," get out of icing conditions as soon as possible and, if ice begins to melt off the airframe, delay landing until the ice is completely removed. If forced to land with ice on the airframe, choose a long runway (you'll be landing fast), don't use flaps, and pick a speed that "feels" comfortable, with a healthy margin above your normal approach speed. This will require more power than usual. Maintain power until just inches above the runway because you might well stall with even a small power reduction. If at any time you feel the beginning of a stall (remember, stall warning devices might not predict the loss of lift in this case), immediately add power and airspeed, and hold that new value until landing.

Many of the ice-related accidents involved induction system, or carburetor, icing. Carb ice can form at even quite warm outside air temperatures, when sufficient moisture in the air freezes in the cooling venturi effect of the carburetor. Take a good look at your airplane's *Pilot's Operating Handbook*, and the engine manufacturer's manual as well if it's available, to learn the recommended use of carburetor heat for the airplane you're flying. It might differ among airplane types. Finally, be watchful for any unexpected loss of power after engine start, and immediately apply full carburetor heat before a restrictive block of ice forms in the induction system.

Initiated/continued IFR flight into known adverse weather

In these cases, the pilot began or continued an instrument flight into areas with reported thunderstorms, turbulence, or ice. The specific hazard encountered was not reported in the raw data I used in my research. Sixty-seven crashes of this sort took place during the study period, and 76 percent of those were fatal. See Table 12-11.

Table 12-11. Initiated/continued flight into known adverse weather

Accident cause	Accidents	Fatal accidents
Initiated IFR Flight into Reported Adverse Weather	31	25
Continued IFR Flight into Reported Adverse Weather	29	20
Attempted Flight into Known Adverse Weather: Alcohol Impaired Pilot	7	6

This is another instance when a pilot either did not choose to believe a weather briefing before takeoff, ignored conditions en route, or believed that he or she could "handle" flight in thunderstorms, heavy turbulence, or ice. Usually, the pilot could not.

Thunderstorms

There were 66 other accidents where thunderstorm penetration was specifically mentioned as a contributing factor in the accident. Eighty-nine percent of these crashes killed somebody, as noted in Table 12-12.

Table 12-12. Thunderstorms

Accident cause	Accidents	Fatal accidents
Structural Failure	32	32
Loss of Control	27	22
Disappeared Overwater: Thunderstorms Along Route	5	5
Landed Short in Thunderstorm/Gusts	1	0
Lightning Strike	1	0

What more is there to say? Know what conditions cause thunderstorms to form, obtain frequent weather briefings before and during flight, and maintain a healthy respect for the forces generated by even a "small" thunderstorm. In its Beechcraft Bonanza/Debonair Safety Review, the Aircraft Owners and Pilots Association Air Safety Foundation makes an interesting observation. Although thunderstorm penetration was among the greatest causes of accidents in the Beech Bonanza in the years 1982 through 1990, none of those accidents took place after 1986. This roughly coincides with the widespread introduction of airborne lightning-strike detection equipment ("Stormscope" and "Strikefinder," to cite brand names). To me, this correlation demonstrates the value of airborne lightning detection as a valid safety device. I encourage pilots of all airplane types to consider adding such a device to the panel of any cross-country capable airplane.

Loss of control: IFR pilot on IFR flight in IMC

This category of accident includes en route control loss by an appropriately rated and current instrument pilot, in a perfectly functioning, IFR-equipped and certified airplane, on an IFR flight plan that was not specifically noted to be in thunderstorms, ice, or turbulence in the NTSB accident report. In other words, these accidents happened when a pilot did everything right up to the point that some sort of cockpit distraction created a loss of control. Thirty-seven accidents of this type were noted in the study years. Eighty-one percent of these mishaps were fatal. Take a look at Table 12-13.

Table 12-13. Loss of control: IFR-rated pilot on IFR flight in IMC

Accident cause	Accidents	Fatal accidents
Airspeed Not Maintained	24	21
Unusual Attitude: Inflight Break-Up	10	6
Loss of Control: Lack of Understanding of HSI/Autopilot	3	3

These accidents are further testimony to the need for regular, recurrent instrument instruction, as well as the requirement for a thorough systems checkout, including autopilot and navigation equipment, when transitioning to a new type of aircraft.

Loss of control: pneumatic instrument failure

In most general aviation airplanes, the attitude indicator and heading indicator instruments are powered by some type of pneumatic system. In many well-equipped types, the attitude instrument might be the only device powered by a vacuum or pressure pump. When the pneumatic system fails, crucial instruments go with it, and the pilot is forced to rely on "partial panel" flying skills that might be long dormant. Sometimes the instrument itself fails while the instrument air pressure gauge reads normal.

In the years 1982–1988, 20 accidents were attributed to the failure of pneumatically-driven flight instruments. Obviously, there must have been numerous instances when a pilot successfully completed a flight without these vital gauges. I know I have, twice. However, every time a pilot crashed after pneumatic system or instrument failure, the accident was fatal. What does that tell you? If you haven't logged an hour or so of partial-panel, basic attitude flight and approaches in the last year, it's probably been too long. Remember that virtually all autopilots and flight directors fail with the malfunction of the attitude indicator or the heading indicator, so you can't rely on "George" to get you out of trouble. Be sure to practice a few partial-panel unusual attitude recoveries as well, just in case things start to get away from you in an actual emergency.

Get some simulator training if possible. One of the "niceties" of being a simulator instructor is that I've been able to present this sort of failure to pilots, usually for the first time, in a realistic setting. Unlike training in airplanes, where an instructor reaches over and covers an instrument to begin partial-panel flight, in the "real world," a pneumatic instrument failure occurs gradually. The pilot, trained to maintain heading and attitude indications, begins to follow the wobbling instruments until pitch and bank excursions can be great enough to incite an unusual attitude. He or she is the one that must identify the failed instruments and make the transition to partial-panel flight. Until you've learned to identify and compensate for failed pneumatic instruments yourself in a real-world setting, you haven't truly learned safe partial-panel flight.

Miscellaneous

The final category of airplane accidents during the study years are in the "miscellaneous" category; there were only six of them, and only one of those was fatal. Table 12-14 lists these causes.

Table 12-14. Miscellaneous weather-related accidents

Accident cause	Accidents	Fatal accidents
Improper Use of Autopilot	3	0
Improper High Density Altitude Operation (in Cruise)	2	1
Improper/Inadequate FSS Briefing	1	0

Again, we have a call to do airplane-specific systems checkouts, to teach proper autopilot operation, and to train in techniques associated with high-density altitude performance.

There were many other density altitude-related accidents in the 1980s, but this was the only one identified in cruise flight by the NTSB, and it was consequently grouped in the "pilot error-cruise flight-weather" category in the AOPA study. Finally, although many pilots might complain that the services they receive from official weather briefers might not be realistic, to Flight Service's credit, there was only one instance in this seven-year period when a FSS employee was faulted for an inadequate briefing that led to an accident. The "weather folks" really do know what they're doing. It's up to pilots to take the information they're given and then apply their own knowledge and experience to safely complete a flight.

Look again at the most common causes of weather-related airplane accidents, ranked by total number of accidents, as shown in Table 12-15. Now let's list the accident categories in order of the most frequent causes of fatal airplane mishaps. Compare Table 12-16 to Table 12-15.

Table 12-15. General aviation weather-related accident cause categories, 1982–1988

1. Turbulence/Wind
2. VFR in IMC (Pilot Rating Unspecified)
3. VFR into IMC (Pilot Not Instrument Rated)

Table 12-15. Continued

4. Improper VFR Altitude or Procedure
5. Landing in Marginal or IMC Weather
6. Ice
7. Initiated/Continued IFR Flight into Known Adverse Weather
8. Thunderstorms
9. Loss of Control: IFR Pilot on IFR Flight in IMC
10. Loss of Control: Pneumatic Instrument Failure
11. Miscellaneous

Table 12-16. General aviation weather-related fatal accident cause categories, 1982–1988

1. VFR in IMC (Pilot Rating Unspecified)
2. VFR into IMC (Pilot Not Instrument Rated)
3. Improper IFR Altitude or Procedure
4. Thunderstorms
5. Initiated/Continued IFR Flight into Known Adverse Weather
6. Loss of Control: IFR Pilot on IFR Flight in IMC
7. Loss of Control: Pneumatic Instrument Failure
8. Turbulence/Wind
9. Ice
10. Landing in Marginal or IMC Weather
11. Miscellaneous

Which categories of accidents most commonly result in fatalities? Table 12-17 ranks accidents by the percentage of total that were fatal. You might be thinking, "I fly a twin-engine airplane, always on an instrument flight plan. What are the most probable weather-related accident causes for me?" Or you might be a Piper Cherokee owner, concerned mainly with what leads to the most accidents in your airplane class. Tables 12-18, 12-19, and 12-20, respectively, list weather-related accident causes for the three airplane breakdowns (single-engine fixed gear, single-engine retractable gear, and multiengine), ranked in order of the total number of airplane accidents. Remember that percentages have been rounded off, and they might not exactly equal 100 percent.

Table 12-17. Percentage of general aviation weather-related causes that were fatal, 1982–1988

1. Loss of Control: Pneumatic Instrument Failure	(100%)
2. Thunderstorms	(89%)
3. VFR into IMC (Pilot Not Instrument Rated)	(85%)

4. Loss of Control: IFR Pilot on IFR Flight in IMC (81%)
5. VFR in IMC (Pilot Rating Not Specified) (80%)
6. Initiated/Continued IFR Flight into Known Adverse Weather (76%)
7. Improper IFR Altitude or Procedure (72%)
8. Miscellaneous (17%)
9. Ice (16%)
10. Landing in Marginal or IMC Weather (4%)
11. Turbulence/Wind (1%)

Table 12-18. General aviation weather-related accident causes: single-engine fixed-gear airplanes, 1982–1988

Accident cause	Accidents	% of total
1. Turbulence/Wind	890	58.6
2. VFR in IMC (Pilot Rating Not Specified)	305	20.0
3. VFR into IMC (Pilot Not Instrument Rated)	126	8.3
4. Landing in Marginal or IMC Weather	70	4.6
5. Improper IFR Altitude or Procedure	47	3.1
6. Initiated/Continued IFR Flight into Known Adverse Weather	28	1.8
7. Ice	21	1.4
8. Thunderstorms	17	1.1
9. Loss of Control: IFR Pilot on IFR Flight in IMC	8	0.5
10. Loss of Control: Pneumatic Instrument Failure	5	0.3
11. Miscellaneous	2	0.1
Total	1519	

Table 12-19. General aviation weather-related accident causes: single-engine retractable-gear airplanes, 1982–1988

Accident cause	Accidents	% of total
1. Turbulence/Wind	146	25.0
2. VFR in IMC (Pilot Rating Not Specified)	133	22.7
3. VFR into IMC (Pilot Not Instrument Rated)	110	18.8
4. Improper IFR Altitude or Procedure	83	14.2
5. Thunderstorms	30	5.1
6. Initiated/Continued IFR Flight into Known Adverse Weather	27	4.6
7. Ice	19	3.2
8. Landing in Marginal or IMC Weather	13	2.2
9. Loss of Control: IFR Pilot on IFR Flight in IMC	12	2.0
10. Loss of Control: Pneumatic Instrument Failure	12	2.0
11. Miscellaneous	1	0.2
Total	586	

Table 12-20. General aviation weather-related accident causes: multiengine airplanes, 1982–1988

Accident cause	Accidents	% of total
1. Improper IFR Altitude or Procedure	102	31.3
2. VFR in IMC (Pilot Rating Not Specified)	62	19.0
3. Turbulence/Wind	50	15.3
4. Ice	28	8.6
5. Thunderstorms	19	5.8
6. Loss of Control: IFR Pilot on IFR Flight in IMC	17	5.2
7. VFR into IMC (Pilot Not Instrument Rated)	17	5.2
8. Landing in Marginal or IMC Weather	13	4.0
9. Initiated/Continued IFR Flight into Known Adverse Weather	12	3.7
10. Loss of Control: Pneumatic Instrument Failure	3	0.9
11. Miscellaneous	3	0.9
Total	326	

Listing the weather-related accident causes by aircraft class (single-engine fixed gear, single-engine retractable gear, and multiengine) allows for an interesting comparison. Table 12-21 lists the accident cause categories, and then the position, or ranking, that those categories took for the different airplane classes. High rankings, and any significant difference in ranking of accident causes, point out specific situations that should concern pilots of those classes of airplane. Let's evaluate what this teaches us.

Table 12-21. General aviation weather-related cause category rankings by aircraft class, 1982–1988

Accident cause	SEF	SER	ME
Turbulence/Wind	1	1	3
VFR in IMC (Pilot Rating Unspecified)	2	2	2
VFR into IMC (Pilot Not Instrument Rated)	3	3	7
Landing in Marginal or IMC Weather	4	8	8
Improper IFR Altitude or Procedure	5	4	1
Initiated/Continued IFR Flight into Known Adverse Weather	6	6	9
Ice	7	7	4
Thunderstorms	8	5	5
Loss of Control: IFR Pilot on IFR Flight in IMC	9	9	6
Loss of Control: Pneumatic Instrument Failure	10	10	10
Miscellaneous	11	11	11

SEF=Single-engine fixed gear airplanes
SER=Single-engine retractable gear airplanes
ME=Multiengine airplanes

Accidents attributed to turbulence or adverse wind are pretty universally problematic to general aviation pilots, only slightly more so in single-engine airplanes. This might be due to the typically higher experience level of twin-engine pilots, but even in multiengine types the incidence of turbulence and wind-related accidents is very high.

VFR in IMC accidents where the pilot is (probably) instrument-rated is universally high among all three airplane classes. It seems that this sort of "poor-decision-making" accident is just as likely to bring down a Seneca or Baron as it is a Skylane or Warrior.

A big difference (ranking of 7 as opposed to 3 for all single-engine designs) exists between singles and twins in the "VFR into IMC (pilot not instrument rated)" category. This suggests that pilots of multiengine airplanes are typically instrument rated, and therefore this type of accident occurs infrequently. Before multiengine pilots (like me) get overconfident, remember what I've just said about "pilot rating unspecified" accidents. Read on for more details about multiengine IMC safety.

Pilots of single-engine, fixed-gear airplanes were much more likely to have been involved in a "landing in marginal or IMC weather" type of accident. Remember that this category was weighted heavily towards attempted off-airport landings when faced with threatening weather. My theory is that single-engine, fixed-gear airplanes are more likely to be equipped only for VFR operations, and their pilots are more frequently not instrument-rated, hence the decision to "put down" in a field when faced with a greater range of weather hazards. In other words, fog or low clouds were more likely to necessitate a forced landing in the single-engine, fixed-gear designs. It's a gross overgeneralization, but cross-country experience levels would be typically lower also, meaning that the pilot might be less capable at in-flight decision-making and more prone to get into trouble. Finally, as a former Cessna 120 owner, I know that there's a bit of the following attitude among some pilots of very small airplanes: "I'll try to get there. My airplane lands slowly enough that I can always put it in a field if I have to."

The higher-performance, retractable-gear and multiengine airplanes shared a much lower ranking in this category. They experienced many fewer off-airport landing crashes (accidents numbered 54 single-engine fixed-gear, 4 single-engine retractable-gear, and 1 multiengine). Instead, the "retracts" and twins more commonly had difficulty landing out of a properly executed instrument approach, or at night in marginal weather conditions.

The number one cause of weather-related accidents in multiengine airplanes was an "improper IFR altitude or procedure." This sort of accident ranks much lower in single-engine designs. It might be that, armed with all the real and presumed safety and redundancy of an extra engine and all its associated systems, a multiengine pilot has more of a "go" mind-set and attempts approaches in conditions that single-engine pilots try to avoid. It might be that "get it there at all costs" pressures of charter operations, corporate flying, and high-end business pilots make going "just a little bit lower" more enticing to twin types. (Passenger flights are limited in IMC by Part 135 of the regulations to multiengine airplanes only.) This is not meant to condemn multiengine airplanes or pilots, as the figures in single-engine fixed and retractable-gear airplanes were still quite high. Instead, this points out that airplane capability might serve to impair decision-making, making the pilot think he or she can tempt the fates with greater success. Historically, that's not true.

On the other hand, the presumed greater experience level of multiengine pilots does in fact show in the rankings for "initiated/continued IFR flight into known weather hazards." Unlike single-engine fixed- and retractable-gear airplanes, both ranking six in this category, this judgment-related cause was number nine in multiengine airplanes. To play devil's advocate, however, one might think that twins are simply less likely to experience the crosswind-related and precautionary landing accidents so common in single-engine designs, and that, removing those categories, twins would fall into line with singles' rankings for other hazards. Remember also that this category excludes those accidents specifically attributed to turbulence, thunderstorms, or ice. Most multiengine pilots are instrument rated, and their airplanes are IFR equipped, and they are therefore less likely to come to grief in low visibility and unspecified weather hazards. My personal feeling is that a multiengine pilot's typically superior experience level does in fact make him or her a little less likely to take off into glaring weather hazards, and that the twin's added capability makes it a little easier for the pilot to escape adverse conditions once they are encountered. In my opinion, however, the benefits of an extra engine from a weather-related accident avoidance standpoint cannot be statistically demonstrated.

Further evidence of this stance comes from the rankings of accidents attributed to ice. While single-engine fixed- and retractable-gear airplanes both rank seven in this category, ice is the number four cause of weather-related accidents in twins. Why is this? Again, I feel this is due to clouded pilot judgment brought on by the real or presumed superior capability of twins. This is probably also influenced by the number of twin-engine designs that are certified for "known icing," which might lead pilots to continue in conditions beyond the airplane's capability. Remember that the definition of even "light" icing includes the statement that continued flight for more than one hour even in icing-certified airplanes might become hazardous, due to ice buildup on unprotected parts of the airframe. Consider that most single-engine airplane engines are not fuel injected and are therefore susceptible to carburetor icing. Most twins are fuel injected, so the "ice" category is even more heavily weighted towards the multiengine pilot.

Taking the risk that you might think I'm unduly "picking on" pilots of high-performance single-engine and twin-engine airplanes, the next cause category, thunderstorms, was also most likely to lead to an accident in those types of airplanes. Ranking eight in the "stiff-legged" airplanes, thunderstorms shared a ranking of five in all other designs. Why might this be? Heavy singles and twins are more commonly used for cross-country and weather flying and are therefore more likely to be exposed to the hazard. I think that this is another testament to the value of airborne weather-avoidance equipment, including lightning detection devices and airborne radar. The rates of thunderstorm-penetration accidents in the more commonly weather-avoidance-equipped, "high-performance" airplanes is becoming demonstrably lower. As cited earlier, there was not a single thunderstorm-related accident in a model 33 or 36 Beech Bonanza from 1986 through 1990.

Again supporting the hypothesis that multiengine pilots are lulled into overconfidence by the capability of their airplanes, a "loss on control: IFR pilot on IFR flight in IMC" was historically ranked much higher (number six) in twins than in single-engine airplanes (number nine for both fixed and retractable-gear models). Maybe the twin-engine pilot should consider the off-airport precautionary landing more frequently if conditions become critical. The more capable the airplane, the more potential distractions it presents, and cockpit distraction is what leads to a loss of control. This is another call for recurrent pilot training, including a simulation of pilot, ATC, and emergency-induced distractions that force a pilot to remember that "fly the airplane" is always first on their list of priorities.

To complete the list of accident cause rankings, pneumatic instrument failures and "miscellaneous" weather-related accident causes were just as likely in any class of airplane. This does raise the interesting point that many single-engine retractable designs have a back-up source of pneumatic air, and most twins have completely redundant pneumatic air sources (one on each engine). Either pilots are not properly checking and maintaining their pneumatic systems, or the pneumatic instruments themselves are breaking regardless of the air flowing across their gyros.

Regardless, pilots need to practice realistic transitions to partial-panel flight. They also need to practice sufficient en route, descent, and approach procedures to be comfortable in avoiding this always-fatal type of accident.

CONCLUSIONS

Historically, weather-related accidents account for 15 percent of all general aviation mishaps. The weather-related accident record is worst in single-engine, retractable-gear airplanes. Nineteen point nine percent, nearly one in five, of all accidents were in part due to aviation weather hazards. Multiengine airplanes have the next highest incidence of weather-related accidents, at 16.7 percent, while one in seven (13.5 percent) single-engine, fixed-gear airplane mishaps were due at least in part to meteorological factors.

When critically examining the specific causes of weather-related crashes, we find that the accidents can almost always be grouped into two human-factors categories: pilot judgment and pilot attitude. In a very large percentage of the accidents, had the pilot obtained a weather briefing for flight, or believed what the forecasters warned, the accident quite likely could have been avoided. An appalling number of accidents happened when pilots deliberately launched into hazards beyond the limits of their ratings, currencies, capabilities of their airplanes, with predictable results. In other cases, pilots demonstrated a lack of understanding of weather development. They continued on in the apparent hope that things would get better, then conditions became beyond their ability to control. It's been said that actual weather is always better or worse, but not precisely as forecast. Pilots need to observe conditions to verify or refute the official weather forecasts, and they need to repeatedly reexamine their decision to continue on or divert in flight. These are all categories of pilot judgment, and I think that the best way to develop weather judgment is to write down precise weather situations or phenomena that call for an immediate "no-go" or diversion decision. Then make a vow to never violate those "personal minimums" because "I've got to make it to this meeting," or "My spouse expects me home by tonight," or "I can't afford to take the airlines home and then come back after my airplane." Work with an experienced instructor who will honestly evaluate your current level of skill to derive your list of minimums, and then expand your personal weather-flying envelope gradually to increase the safety and utility of personal flight.

Pilot attitudes are also evident in accident statistics. I don't think it's a coincidence that high-performance, single-engine airplanes have the worst weather-flying record. It's easy to get overconfident in a Mooney or a Bonanza, simply because these fine airplanes are so capable. Add another engine and the probability of weather radar and ice protection equipment, and you'll see how multiengine pilots can be lulled into a false sense of security about weather hazards. What causes the "high-end" singles and twins to crash, however? Turbulence, crosswinds, busting minimums, and loss of control in the clouds (the same sort of conditions that bring down fixed-gear singles except that the rate is typically lower in the single-engine, fixed gear airplanes in part because their pilots know to avoid these sorts of hazards). Think about my own experience, which I related earlier, where switching from a single-engine, retractable-gear Bellanca SuperViking to a fixed-gear Cessna Skyhawk made me suddenly so concerned about the development of ice. High-performance airplanes can lead even experienced pilots into believing they are more capable than they really are.

How do we educate ourselves about our own attitudes? Precisely define the weather hazards you and your airplane are capable of handling physically, mentally, and legally. If weather conditions violate those defined limitations, don't go. Divert or land as soon as practical. Convince yourself that aviation training isn't an isolated event, but is instead a continuing process. When was the last time you took some flight instruction? Is this the first aviation publication you've read recently that describes techniques for dealing with weather hazards? Do you regularly practice unfamiliar approaches to minimums, or low-speed, low-altitude maneuvering, or partial-panel instrument flight? Unless you fly every day, in varying weather conditions and over long and diverse routes, it's virtually impossible to stay current based on experience alone. That's where training, which is in reality merely learning from the experience of others, comes into play. Some might look on flight instruction as an admission of lack of skill, but flying is like learning a foreign language. If you stop speaking it regularly, you'll eventually forget all but a few phrases. Put another way, would you want a surgeon to operate on your heart if he or she had been taught to minimum standards and had not practiced the procedure in five years? That's precisely the sort of life-threatening risk a pilot who doesn't train regularly takes. Think about the unknowing passengers who fly along with that pilot.

General aviation is enjoying a continuous string of its safest years in history. As airplanes and navigational equipment become more and more sophisticated and reliable, however, accidents brought on by lapses in pilot judgment or attitude have come to the forefront of causes of mishaps. Nowhere is this more apparent than in general aviation weather-related accidents.

PART 3

Regional weather

13 Regional weather patterns

THERE'S NO SUBSTITUTE FOR EXPERIENCE IN FLYING IN LOCAL WEATHER patterns. If you're flying in unfamiliar territory, the next best thing to experience is talking to someone who's intimately familiar with the local meteorology. Short of that, however, it is possible to anticipate typical local weather patterns if you know the general rules of weather development and something about the regional topography. In this chapter, we'll look at the general weather flow across the continental United States, and then we'll break this down into eleven regions, considering regional terrain to derive the basic rules of thumb for each region's weather.

THE CONTINENTAL UNITED STATES

We've already seen much of what the topography of the United States, on a continental scale, does to patterns of weather development. Inland areas are generally dry and colder further north; coastal areas are warm and moist by comparison. The low-pressure areas that tend to form inland, as well as the prevailing southwest-to-northeast circulation of the atmosphere at these latitudes, pull moisture from the Pacific

Ocean and the Gulf of Mexico to combine with cold air inland from the north. To a lesser extent, water from the Great Lakes and the Gulf of California also affect continental weather patterns. A semipermanent low-pressure system in the North Pacific pushes moisture in to create adverse weather (the water moderates temperatures as compared to arctic-influenced air inland); the semipermanent Bermuda High, (air colder than semitropical areas ashore) halts advancing fronts in the American southeast or pushes them up the Eastern Seaboard (Fig. 13-1). Let's look now at some specific regions of the United States to see what weather patterns typically occur.

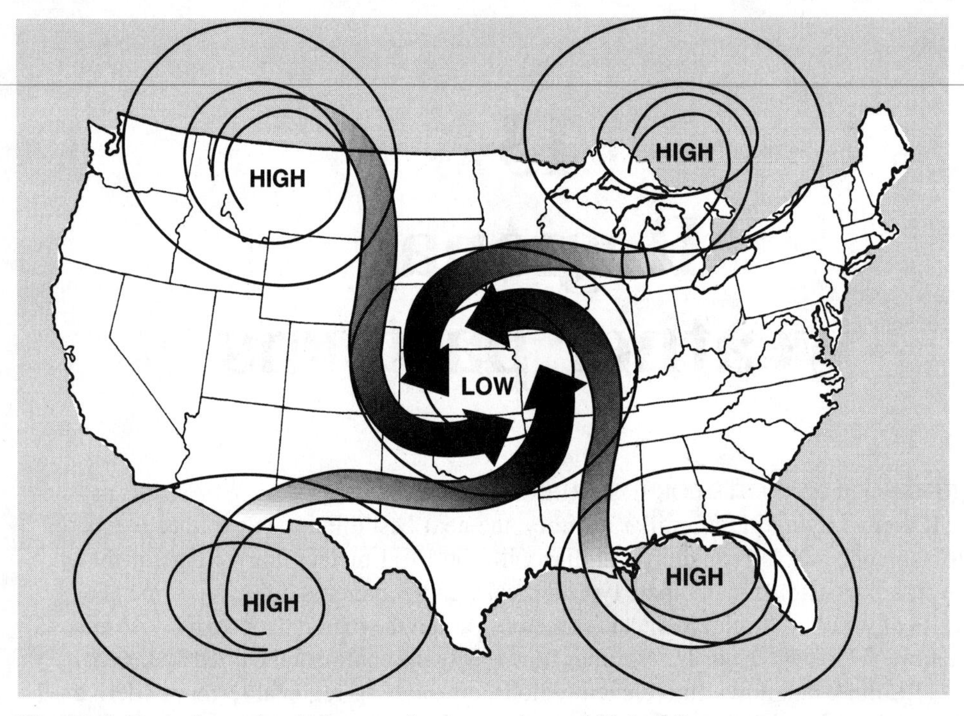

Fig. 13-1. *Typical weather influences in the continental United States.*

The Pacific Northwest

The Pacific Northwest, defined as the area of Washington, Oregon, and northern California, is known for its moist, moderate climate. In fact, the Puget Sound area near Seattle is the wettest area on the American mainland, with an average of 150 inches of rain annually; this area is the home of the only nontropical rain forest on the planet.

This region is characterized by proximity to the Northern Pacific low and a spine of tall mountains aligned north-and-south with few passes (Fig. 13-2). In summer months, when inland areas warm, the Pacific low translates north into the Gulf of

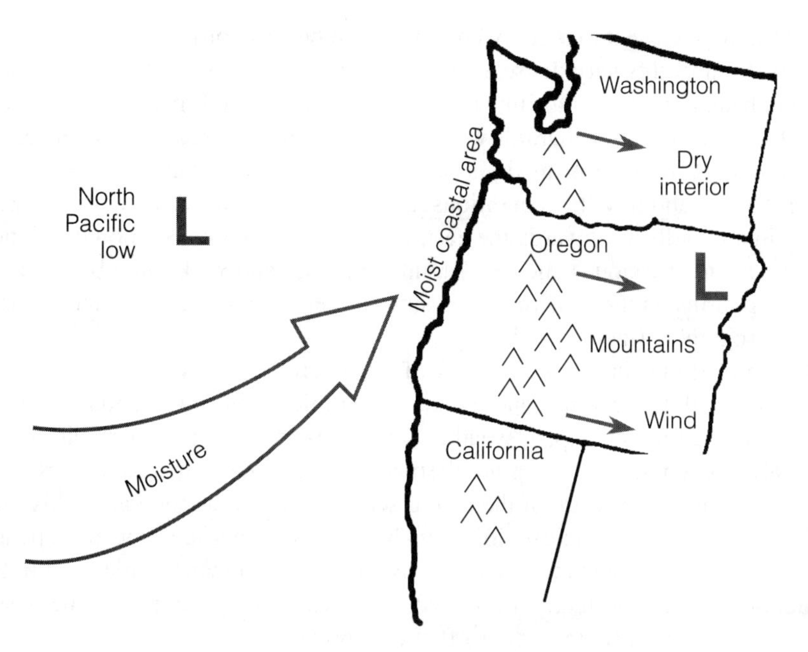

Fig. 13-2. *Typical weather influences in the Pacific Northwest.*

Alaska, pushing huge amounts of moisture into the region with a succession of weak warm and cold fronts. In winter, this climatological feature moves southward, making the southern part of the region wetter than "up north." Consequently, without a big, windy weather system to upset the "natural" state, the windward side of the Cascade Mountains is foggy and rainy. IMC conditions exist 50 percent of the time and improve on average in summer. Clouds are thick with moisture, but generally smooth for lack of convection. I remember an airline trip into Seattle where the clouds were so thick I couldn't see the airplane's wingtips on approach. Because of the moderating influence of all this water, however, the inland areas are cool but not cold year-round, and this moisture is usually in the form of rain, not snow at the lower elevations.

Conversely, east of the mountain range, the Pacific Northwest states are arid desert. Places like Yakima and Spokane average less than 10 inches of rain annually and suffer wide temperature variations, scorching summers with arctic winters.

The movement of the North Pacific low creates two seasons in the coastal Northwest. Winter is the wet season, with rain on average 20 days of every month. If the wind is out of the south, meaning a Pacific front is approaching, expect rain; if from the north, behind a cold frontal passage, severe icing is common due to atmospheric lifting in the Cascades. Thunderstorms are relatively rare because of the lack of temperature contrasts, but thunderstorms might occur if unstable air blows south from Canada. Mountain-wave turbulence might form east of the Cascades if wind speeds exceed 30 knots at the ridge height; this turbulence can extend hundreds of miles inland. If inland areas get extremely cold, possible because of the lack of moisture east of the mountains, this very dense air might blow downhill towards the coast, accelerating and heating in the venturilike mountain passes to speeds near 80 knots. Wind reported from the east, then, heralds warming, clearer air, but heavy turbulence.

By contrast, summer is the dry season for the coastal Northwest. Dry, however, is a relative term, and low IFR conditions are still common for days at a time. Extreme heating inland pulls air towards the east, air that brings copious amounts of moisture inland to create low clouds and fog. Mountain passes and peaks are obscured, and radiation fog forms on the wet, vegetated slopes. Weak, embedded thunderstorms might form by orographic (mountain) lifting.

If you're flying into coastal areas of the Pacific Northwest, then, expect thick cloud layers with low ceilings and reduced visibilities. VFR pilots should plan several days' extra time for their trip, as weather conditions are often poor. An IFR arrival will almost always require declaring an alternate, which usually means a short trip eastward to clear air, but beware of the lifting action of the high mountains; this region is also known as one of the iciest aloft anywhere, as wet Pacific air pushes upward and cools along the ridges, and mountains in any case usually mean turbulence. Flights east of the mountains are normally VMC, but extremes of temperature and the possibility of mountain waves mean the ride is often a bumpy one.

Northern California

Northern California is another battleground between ocean air and rising terrain (Fig. 13-3). The average temperature hovers near the average dew point, meaning that widespread fogs of long duration are common. In summer, northern California is under the influence of both the cold Alaskan current running down the coastline and the Pacific High, where deep ocean waters keep the air relatively cool. This semipermanent feature pumps a wet northwest wind to the region, making the area quite moist.

Meanwhile, arid areas inland pull this moisture over the cool coastal land, triggering advection fog that advances like clockwork in the late afternoon. I watched the fog roll in daily around 2 p.m. while based at Vandenberg Air Force Base near Santa Barbara; it dissipated around 10 o'clock each morning. This fog starts off as a thin layer, less than a thousand feet thick, which quickly burns off the next morning. With each passing day, however, more moisture is added, creating a thicker layer that blots out the sun for longer periods each day, until finally a heavy fog might linger for days at a time. This heavy fog itself is drawn to inland valleys, cutting off their sunlight and shutting down the "fog machine" until some unrelated weather system blows the moisture away and the inland heating resumes.

In winter, northern California is the recipient of moisture from the North Pacific low. Fog drifts in from the sea with the tide and lingers unless high tide is early enough in the day to permit sufficient heating. Planning a flight to northern California? In winter, ask the National Weather Service for the times of high tide, and plan to arrive or depart before the tidal fog forms. In summer, expect foggy afternoons and evenings; plan

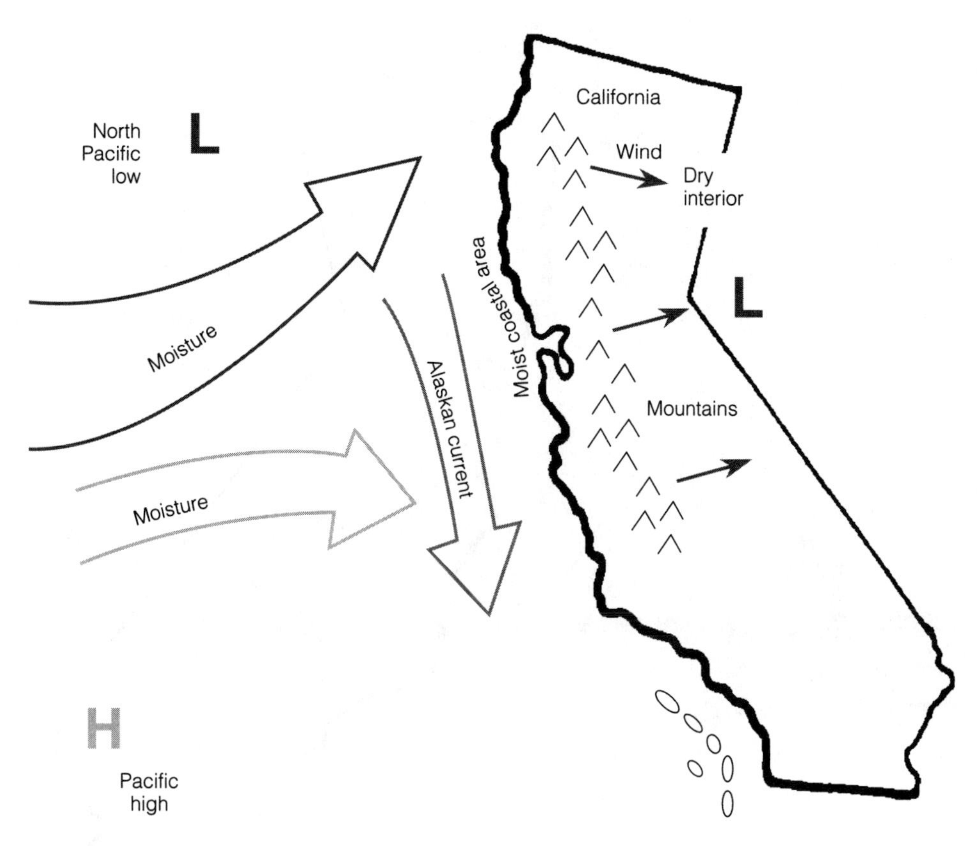

Fig. 13-3. *Typical weather influences in northern California.*

an arrival or departure in late morning or early afternoon, after the fog has burned off but before it reforms. In either case, an alternate is usually available a short distance inland but, if no fronts have blown through the area for several days, even inland valleys can be obscured for long periods of time. Snow and ice might hamper a cross-mountain flight, and mountain-wave turbulence east of the ridges is possible if conditions are right.

Southern California

Southern California has a reputation as a warm, dry place; in fact, the Los Angeles area was originally a desert, now green with irrigated vegetation. This region's weather makers are the cold ocean current flowing south from the Gulf of Alaska, a hot interior desert, and a steep ridge of mountains near the coast (Fig. 13-4). In short, it is another version of the topography found further up the coast, except that it's far enough south that the beach areas are typically warmer than those in northern California.

The normal pattern for southern California, then, is that heating in the desert interior creates a wind flow inland; fog or a low cloud layer blows onshore in late afternoon or early evening. Air from over the cool sea current creates a strong inversion as

Fig. 13-4. *Typical weather influences in southern California.*

it translates inland, trapping this foggy moisture sometimes for days at a time. If winds are strong, the cloud bases are higher, but beware of the rapid rise in terrain to the east; a 1200-foot ceiling at Orange County or LAX could mean a 200-foot overcast and mountain obscuration less than 30 miles inland. Dissipation of these clouds, should temperatures warm, creates a thick haze in the pollution-rich coastal air.

Spring brings more extensive cloudiness, typically, with cloud tops to over 4000 feet that last for days. In summer, there's less of a difference in air temperatures on the coast versus inland, and cloud layers tend to be thinner and less long-lasting. In winter, a strong high pressure (brought on by cold, interior deserts) flows downhill through the passes, accelerating and heating (by compression) into the 80-mile-an-hour Santa Ana wind. The results are severe turbulence near the surface and reduced visibility in blowing dust.

Are you flying to coastal southern California? Expect a cloud layer in late afternoon, especially in the springtime; mountain peaks and passes might be obscured.

Haze might reduce visibilities to near zero, especially when flying into a rising or setting sun, so plan a time or route if possible to avoid pointing at its shiny disc. VFR pilots might consider staying above the haze layer and carefully spiraling down directly over their destination, if traffic and terrain permit, to minimize their time in the reduced visibility near the ground. As an alternate, and for those pilots with a destination east of the coastal mountains, VFR conditions are almost always to be had in the inland deserts, but beware convective turbulence in summer and the Santa Ana hazards in the cold months.

Desert Southwest

This arid region is characterized by turbulence and high-density altitudes. The hot desert air of Arizona, New Mexico, Nevada, and Utah forms a semipermanent low-pressure area, drawing moisture from as far as the North Pacific and the Gulf of Mexico (Fig. 13-5). The ring of mountains around this region lifts the moisture into towering thunderstorms; the northeast corner of New Mexico averages more than 70 thunderstorm days per year, more than any other part of the United States except Florida; because of the high elevations near the slopes, hail is frequent, and microbursts are common beneath the virga that falls from high cloud bases into the dry air.

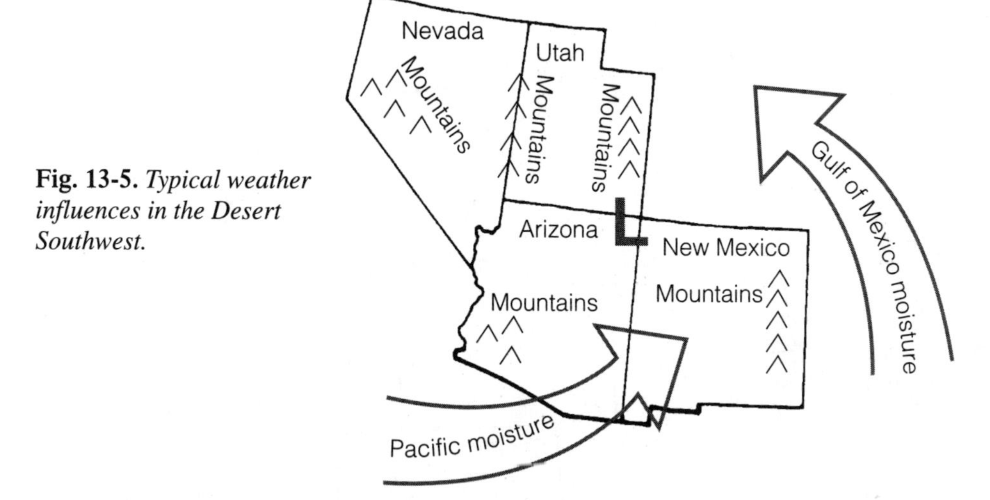

Fig. 13-5. *Typical weather influences in the Desert Southwest.*

The extreme temperatures of this region cause convective turbulence of even severe magnitude up to 15,000 feet, so most pilots try to end flights by noon, before the day's maximum heating. Turbulent air flow around mountains adds to the discomfort of a flight. Scorching temperatures can pull in enough moisture to rain for a week at a time, similar to the monsoons of southeast Asia, but cloud bases are usually high, with VMC conditions beneath. In winter, the strong Pacific low and the equatorial jet stream feed moisture, causing an average 100 inches of snow annually on the ridges and showery rain at lower elevations. Mountain waves are common in the lower jet streams of winter.

Flying to the Desert Southwest? Plan to get up early and fly before noon, especially in summer. Watch for the rapid formation of hail-bearing thunderstorms near peaks late in the day. Snowstorms and icy cloud cover might hamper a winter flight. The most common lament I've heard from pilots living in this spectacular zone is that it's virtually impossible to stay current in instrument flight because 90 percent of the time the weather is "severe clear," and when it is IMC, it's too hazardous to fly.

The Rocky Mountain states

The Rocky Mountain region represents a composite of the weather features of the West Coast and the Desert Southwest (Fig. 13-6). High mountains provide orographic lifting of air, creating thunderstorms, turbulence, and heavy rains and especially snows when moisture blows inland from the west. Rain evaporating as virga in the arid air commonly incites microbursts, especially on the front range of the Rockies in Colorado and Wyoming. Mountain-wave turbulence frequently forms with winter's lowered jet stream flow; cold air in the uplands might descend and accelerate in valleys to become the Chinook wind, raising temperatures dramatically at the mountain bases while at the same time kicking up dust and boiling into severe turbulence in 80-knot winds.

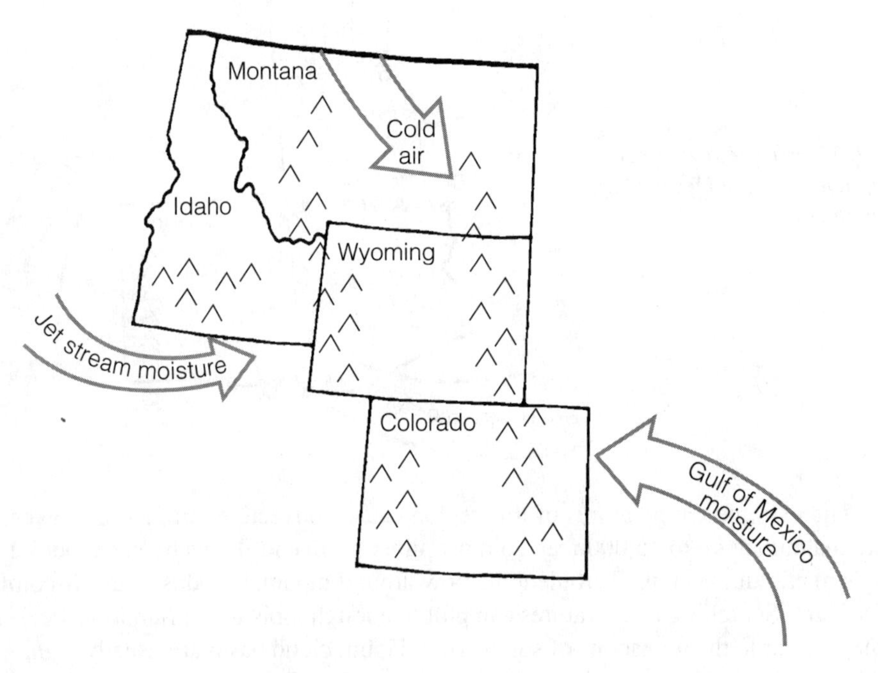

Fig. 13-6. *Typical weather influences in the Rocky Mountain states.*

Don't let this dire-sounding prediction deter you from enjoying a flight in this beautiful region; simply expect unpredictable and rapid changes in the weather if flying in the Rocky Mountain states; before launching alone, it's mandatory to take a mountain flying course in the region you'll travel, and bringing along a local "old salt" as a guide might be a good idea. Review best-power mixture leaning techniques before you fly, as a few airplanes crash each year attempting to take off in high-density altitudes using a "full-rich" mixture setting.

The Great Plains

The flat expanse east of the Rockies and west of the Mississippi is the battleground between Gulf of Mexico moisture and cold air from the Canadian north. The air is comparatively dry, meaning the weather is VMC much of the time, and, with no major obstacles to the passage of air, high wind speeds at the surface are common. I've learned from experience that you can just about double the wind speeds at the surface or aloft when you cross the Missouri/Kansas border westbound (Fig. 13-7).

The Great Plains are known for towering thunderstorms and raging blizzards, both the result of atmospheric lifting mechanisms. Such a large, relatively dry area spawns massive convective heating currents, pushing moisture upward to condensation; an

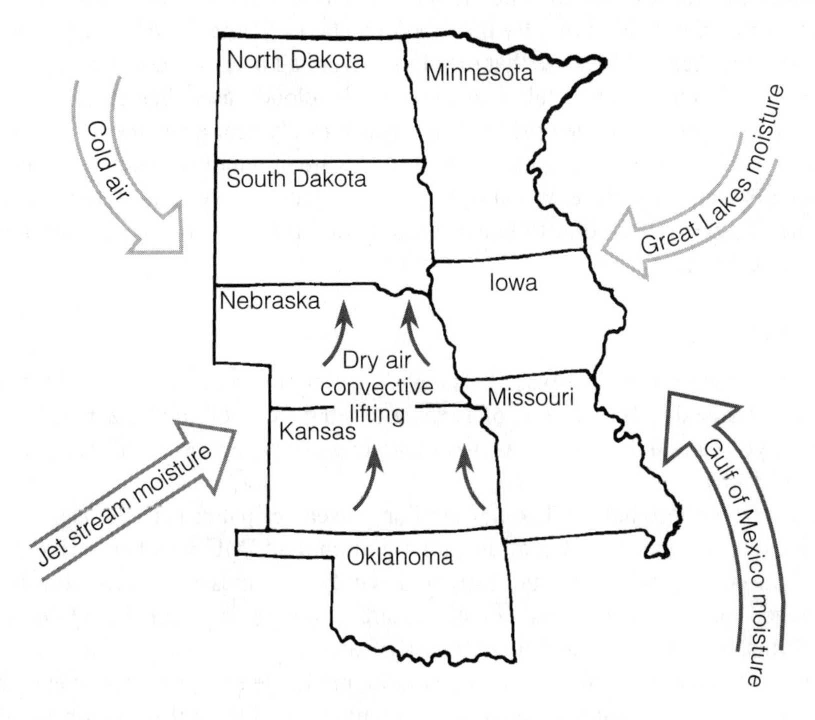

Fig. 13-7. *Typical weather influences in the Great Plains.*

unimpeded jet stream aloft can create suction like a vacuum cleaner to pull wet air upward to develop storm clouds. In winter, frigid air from the north contacts moist air from the south, resulting in heavy snows at the change in seasons. The depths of winter are cold but usually dry, with bone-chilling winds.

The transition to summer, with the greatest contrast between temperatures north versus south, brings the severe storms season. The Great Plains region suffers the most consistently severe weather in the world, as hot, wet air from the Gulf condenses into clouds of extensive vertical development, rising to altitudes above even the tropopause. Tornadoes might form in the muggy air, with spiraling columns of low pressure and winds clocked at more than 250 miles per hour. Summer itself is hot and dry, with the occasional severe storm or frontal passage; autumn, like spring, is a time of contrasts and therefore hazardous weather.

Plot the field elevations from the Mississippi River to the Front Range, and you'll find that the ground gradually slopes up from east to west. If the wind is from the east, moisture from the Great Lakes and the Gulf of Mexico rides upslope and cools by expansion; extensive fogs are common in the midsummer and midwinter lulls. I drove from Wyoming to Wichita one July weekend and saw a great example of upslope fog. Crossing the last ridge before Cheyenne eastbound, I found an undercast that I drove into when descending into Wyoming's capital. The fog persisted until a couple of hundred miles east into Nebraska, when it became a low, scudding layer, gradually lifting as I put more miles behind me. By the time I reached Wichita, I was under a 3000-foot overcast; only then did I realize that the cloud layer had been at a consistent altitude all this time. Only the land had fallen away from the cloud bases. Winds were light and out of the east, providing the upslope flow. You'll rarely hear a weather briefer call out "upslope conditions" for the Great Plains, but beware of winds out of the east in an otherwise calm weather picture. When flying to this region, also watch for rapid storm development, especially in mid to late afternoon, and practice your crosswind landings before making the trip.

Texas

Simply because of its size, Texas hosts a wide variety of weather. In fact, the National Weather Service divides the state of Texas into ten different climatic zones. Basically, however, you can think of Texas as two separate weather regions: a dry half, and a wet half (Fig. 13-8).

The dry, western half of Texas is similar in weather patterns to the Desert Southwest. Extreme temperature variations are common, and IMC weather is rare. The region is known for whistling, unobstructed winds, and massive convective heating. Turbulence and thunderstorms, therefore, are common in summer, while chilling windstorms, ice, and snow are the hallmark of winter.

The wet, eastern half of Texas enjoys more moderate temperatures, evened by the influence of Gulf moisture. Coastal fog is common, and the entire region is IMC 33 percent of the time. Rain and even thunderstorms mar the winter skies, as a semiper-

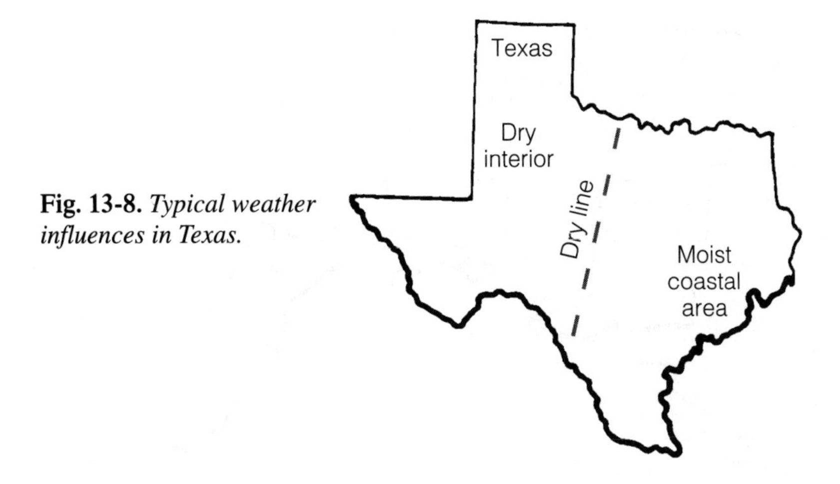

Fig. 13-8. *Typical weather influences in Texas.*

manent Gulf low pumps air towards the colder hill country. Transitions into and from the hot, humid summer are marked by heavy thunderstorms; the Dallas and Houston areas experience severe thunderstorms as much as 60 days of every year, and of course the region is exposed to hurricanes travelling in from the Gulf.

At the conflict of the wet half and the dry half there is a region of unsettled weather. In winter, this big temperature and moisture difference triggers heavy rains, snows, or thunderstorms. In summer the temperature difference isn't enough to form a conventional front, but a "dry line" develops where the dew point drops off rapidly. This is the "Marfa Front" we discussed earlier in this book. Essentially a trough that acts like a sharply defined cold front, the dry line forms in the west, creating a late-morning line of storms that intensify and blow east before dying out in early evening.

Planning a flight to Texas? Expect clear, turbulent air in the west, and muggy, IMC weather in the east. Thunderstorms are common year-round, while ice and snow occur in the higher western elevations, and extensive areas of rain occur towards the coast. Ask your weather briefer about dew points for your route and west of your route of flight; a big drop in a short period of time warns of the dry line and its associated storms. You might want to plan your trip for early in the day or after dark to avoid the worst storm hazards, and remember that a diversion to the east will usually get you away from a rapid line of storms. Hone up your crosswind and wind shear technique if you're flying to western Texas.

The Gulf Coast

As the name suggests, this region is influenced primarily by the moisture of the Gulf of Mexico. A semipermanent high-pressure area over the cool, deep waters near Bermuda pushes air up from the south, bringing high humidities, rain, and fog as this moisture hits cooler air flowing from the north (Fig. 13-9). Humidities along the Gulf

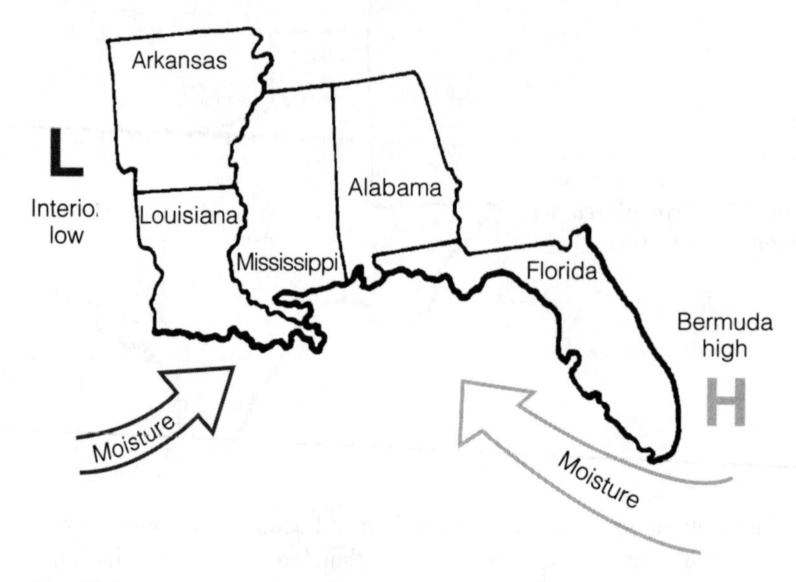

Fig. 13-9. *Typical weather influences along the Gulf Coast.*

Coast average in the 80 percent range, meaning that low clouds and fog are the most common hazards to pilots.

Thunderstorms are common here as well, especially at the change of seasons when there is a strong contrast between cold air inland and the moisture from the waters. Seventy to 100 thunderstorm days a year are typical. Long, drenching rains fill the bayous as well; coastal regions of Louisiana, Mississippi, and Alabama average more than 60 inches of rain a year. The area is, of course, a target for hurricanes in the autumn. Despite this gloomy forecast, however, IMC weather prevails only about 20 percent of the time, and the warming influence of the Gulf keeps these states from experiencing the scorching heat and frigid cold of more inland areas.

Flying to the Gulf Coast? Watch out for morning fogs, especially in winter, and thunderstorm activity in the late afternoons. Beware of cold fronts from the north, or a strong, wet and weather-laden breeze from the south.

The Southeast

This part of the United States is influenced by massive amounts of moisture from both the Atlantic Ocean and the Gulf of Mexico. The Bermuda high circulates wind from the south or southeast, pumping wet air over the land, so beware of any cooling influence from the north; it will create weather hazards such as storms and reduced visibility (Fig. 13-10).

Cold fronts from the northwest tend to stall out in this region, butting against the strong flow off the Bermuda high. The Atlanta area, for instance, is known for the soaking rains or thunderstorms that last for days and are created by this station-

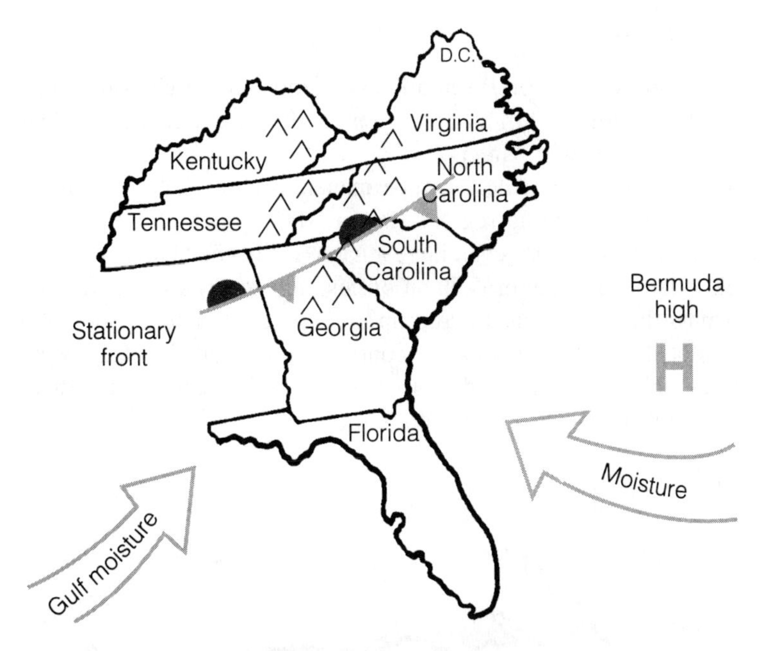

Fig. 13-10. *Typical weather influences in the Southeast.*

ary-front producer. Frontal weather exists in northern Georgia and the Carolinas more than 30 percent of the time, especially in winter. Orographic lifting from the southern spine of the Appalachians fills these stationary fronts with embedded thunderstorms; the Charlotte area especially is known for freezing rain and ice-laden clouds in winter.

Winter months are commonly foggy, as longer nights allow for more radiational cooling and are intensified by a strong wind off the water. The mouth of the Mississippi and other rivers are especially potent fog-makers; cold river waters bring cold air flows with them to condense coastal humidity. Light winds tend to lift this into a low cloud layer. The long heating days of summer lead to thunderstorms along the coasts in summer, especially where aided by the sea breeze along the coast and in Florida. Hurricanes, massive low-pressure systems containing monstrous winds, low visibilities, and thunderstorms, are most common in the autumn months, when the Gulf waters are at their warmest.

Plan your flight anywhere along the coast to take place in the late morning or early afternoon, after the morning fogs dissipate but before storms mount. Expect delays due to ice and the almost constant frontal weather in the north part of this region; be especially watchful for a warm front off the Gulf of Mexico, which almost always means long periods of low weather. Don't discount the possibility of mountain-wave turbulence in the lee of the Appalachians, either; it's actually common in the western Carolinas in the winter months.

The Great Lakes

The Great Lakes provide a significant amount of moisture directly in the path of cold Canadian air. This humidity will moderate temperatures, bringing cool, clear summers but foggy, icy, snow-covered winters.

In summer, the lakes provide a semipermanent high-pressure system, as the air is cooler than that in surrounding regions (Fig. 13-11). This high pressure repels most frontal activity; meteorologists even have a name for it, the "Omega Block," because of the Greek-letter-shaped pattern's habit of pushing stormy weather around the lakes region. Air-mass thunderstorms do crop up in the late afternoons, but they tend to be small and isolated. As the hot months continue, however, the lakes warm and the "Omega Block" breaks down, leading to a period of hot, muggy weather in August when severe storms and tornadoes are common.

Fig. 13-11. *Typical weather influences in the Great Lakes region.*

Winter comes early to the lakes region, but it is not as harsh as other areas of similar latitude because of the lakes. A "lake breeze," similar to sea breezes off the ocean, makes fog common near the shores in the late afternoon and evening. Strong winds from the northwest blow across the lakes, gathering moisture, only to dump it as massive snowstorms when condensing over the cold shores on the lee side of the lakes. These *lake effect snows* are some of the heaviest in the world, limiting visibilities, loading airplanes with ice, and closing airfields. The lake effect is most common in the first half of winter; when the lakes freeze solid, the lake effect no longer occurs.

Expect relatively cool, clear air and fair-weather cumulus clouds when flying over the lakes region in summer, but watch for the rapid development of thunderstorms late in the day. If it's August, temperatures will heat up and the storms will increase in frequency and severity; "tornado alley" actually migrates to the Great Lakes this late in the season.

Winter's biggest threat is the ice and snow, most common east of the lakes. Watch for a stiff wind from the northwest when anticipating rain, ice, and snow in Michigan, Ohio, and New York.

The Northeast

The Northeast's main weather producers are the waters of the North Atlantic, cold, Canadian air from the north, and the influence of the northern Appalachian mountains. The typical route of the jet stream means that weather systems funnel over this area from the western United States (Fig. 13-12).

A semipermanent Atlantic low, over waters relatively warm as compared to inland areas, pumps moisture in from the northeast; if well developed, this is the classic "nor'easter" of whaling lore. This wet air mass contacts cold air from Canada, bring-

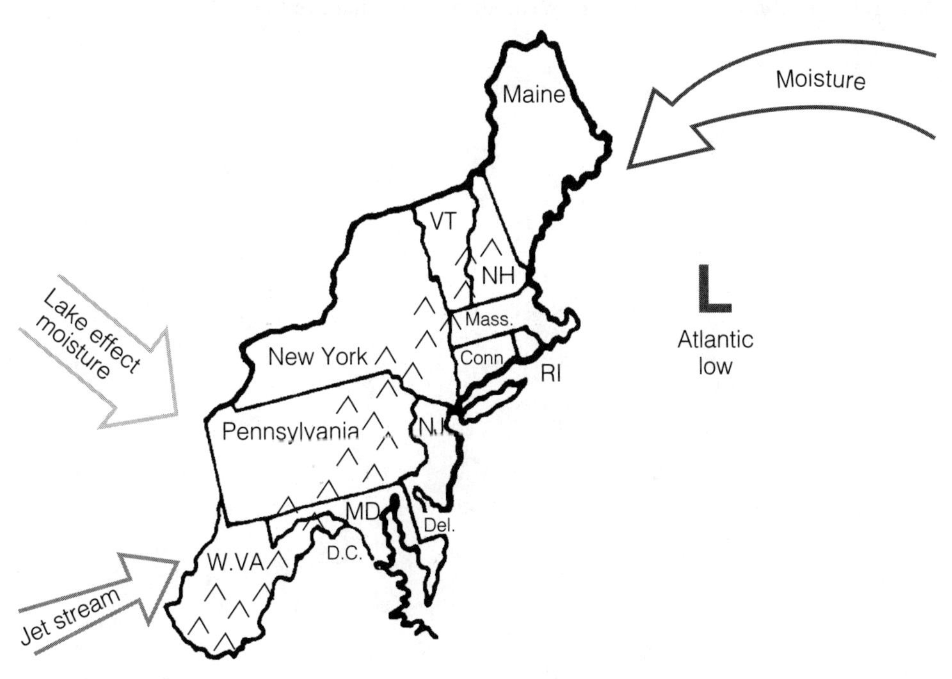

Fig. 13-12. *Typical weather influences in the Northeast.*

ing measurable precipitation in the area every three days on average, and causing IMC weather 15 percent of the time in summer (33 percent in central Massachusetts), 50 percent of the time in fall, 33 percent of the time in spring, and 10 percent of the time in winter. Dense fog from the sea breezes can travel scores of miles inland; the thick layer, with tops commonly to three or four thousand feet, can last for days. Airframe ice is common in the soupy cloud layers, especially in the vicinity of the lifting action of mountains; weak embedded thunderstorms, still dangerous to aircraft, pop up to the east of the mountain ridges.

Flying to the northeast? Brush up on your instrument-flying skills. Not only will this permit you freer passage through the commonly clouded skies, it will also help you cope with the demands of flying in the busier terminal areas. If flying VFR only, budget a lot of time for postponements and reroutings. Watch for the warnings of strong winds aloft: from the east this means worsening weather; from the west it could mean embedded thunderstorms, ice, snow, and mountain-wave turbulence in the lee of the mountains.

These regional weather pictures are of course gross simplifications of the typical patterns region by region; day-to-day weather patterns differ significantly. These scenarios will, however, give you an idea of what to expect if no other significant pattern exists in your preflight briefing. There's no substitute for experience when flying in local weather patterns, but, barring that personal experience or advice from a weather-wise local, you can use the general rules of weather development and your knowledge of terrain to anticipate the types of weather hazards that are likely to occur.

PART 4

Flight planning

14 Making the go/no-go decision

YOU'VE AMASSED A GREAT DEAL OF INFORMATION ABOUT WEATHER THEORY, aviation weather hazards, and regional weather patterns. How can you put this data to use to make a more confident, better-informed go/no-go decision?

GOALS OF THE FLIGHT

Regardless of the endeavor, whether it concerns business, finance, or your personal life, you need two items in order to succeed: a set of goals, or priorities, and information on which to make a decision about how to meet those goals. This holds true in aviation as well. Before you can make a confident and informed decision on whether to make a flight, and the route by which you'll make it, you need to identify your goals.

Think for a minute about why you fly. Make a personal list of the reasons you fly an airplane. Why do you fly instead of drive? Tops on most people's list is time. You save a lot of time by flying. Why don't you go by the airlines? Convenience is the most common answer to that. The airlines often don't go when you want, or for that matter, where you want; valuable business or recreation time is spent driving to and from commercially served airports and waiting for plane changes and luggage pickup. An oft-quoted second reason is comfort. Airline travel in many cases has become an ordeal of long lines, narrow, cramped seats, and sprints between airplanes at distant hub airports.

Why don't you charter an airplane? You avoid the hassles of airline travel and still arrive where you want, when you want. Cost is usually the answer to this; chartering an airplane is not an inexpensive venture. Close behind on most pilots' lists is enjoyment, or satisfaction. There is a great deal of personal satisfaction in piloting an airplane safely to your destination.

There might be some financial incentive to going by air. This is most often true of business people who fly; a personal airplane means serving customers better or beating the competition to a market.

If you listed these priorities, or goals, in piloting a personal airplane, you'd have something like this:

- Time.
- Convenience cost.
- Comfort.
- Enjoyment.
- Satisfaction.
- Business (profit).

You might come up with more.

I mentioned one other goal, one that doesn't make it onto most people's lists when I present this exercise in seminars: safety. "That goes without saying," you might reply, but in the experience of many pilots and the National Transportation Safety Board, it does not. How many times have you read of a pilot flying VFR into IMC conditions and crashing, in many cases even when the pilot was instrument-rated? That's a classic case of inappropriate priorities, putting one goal (probably time or cost) ahead of another more important goal—safety. What about the pilot who leans the mixture a little richer than "book" procedure ("Running the engine cooler lengthens its service life," an arguable position), only to run out of fuel short of his destination because the airplane couldn't meet its *Pilot's Operating Handbook* endurance? This is another instance when the goal of lowering costs is put ahead of safety.

Refigure your goals with safety in mind, and you'll see that safety should be number one on your list of priorities. All other goals are interchangeable. You might sacrifice a little cost, in terms of fuel burn and engine life, by using a higher power setting to gain time, for instance, but no goal should be sacrificed for safety. Taken to the extreme, this would mean never flying (or getting out of bed, for that matter), so what you need to do is to prioritize your goals and use them to manage risk, to reduce it as much as possible.

Why do I spend so much time on goal-setting in a book on aviation weather? It's because pilots historically do a very bad job of risk management when weather is concerned. Remember, approximately 25 percent of all general aviation accidents, and nearly 40 percent of the fatal general aviation accidents, have historically been caused by weather factors. In most of these cases, pilots either did not obtain a weather briefing prior to flight, chose to ignore warnings about weather hazards that they did get from briefings, or failed to react to changing weather conditions en route soon enough to prevent disaster. In order to meet your primary goal, safety, you need to factor in the weather when determining how a flight will progress.

PLANNING THE FLIGHT

In including weather in your go/no-go decision-making, you need to reevaluate your goals for a flight. Let's say, for instance, that you have a business meeting in Minneapolis, and you're currently in Wichita, Kansas. Consciously reviewing your goals, you decide that they are:

- Safety: always first!
- Time: you'll save days compared to driving, and you'll save several hours compared to flying a circuitous airline route.
- Convenience: you avoid the hassle of driving to and from air carrier airports, and you dodge the logjam of a hub airport en route.
- Business: there's profit to be had in Minnesota!
- Enjoyment: the airline super-saver fare might be cheaper, but face it; flying's a lot of fun!

You need to decide whether or not to go, and the route you'll take to get there, to meet your goals.

Look at the weather. You want to pick a route that represents a minimum of risk. Evaluate the weather briefings your receive on the basis of the four categories of hazard (thunderstorms, turbulence, reduced visibility, and ice), and see which altitude and route exposes you to the least of those hazards. Always have in mind an escape route. If the turbulence becomes too great, or thunderstorms block your passage, what is your way out? If you know beforehand that, for instance, a turn eastward means improving weather conditions, you'll have much less stress (read, the ability to make a better decision), if a diversion becomes necessary. If ice begins to form on your windows and wings, you should know whether a climb or descent will put you in warmer air, or which direction will bring improvement. If your destination goes below minimums, or an emergency forces you to divert along the way, you should already know where the closest marginal VFR and VFR weather prevails. When ATC asks, "What are your intentions?," you've already made the decision; just back it up with the latest information and do it.

As a way of backing up safety as a primary goal, many pilots employ "personal minimums" for ceiling, visibility, and crosswind component. Personal minimums should be based on recent experience. Just because you passed an instrument competency check six months ago, or landed your airplane in a 20-knot crosswind last April, that doesn't mean you're proficient in the task now. The airlines require pilots with limited experience to increase minimums for a landing. I was on a 737 once that missed the approach even though the ceiling was 400 feet, with three-quarter-mile visibility; that airline's operating rules specified higher minimums for a new 737 captain, which ours was, and we landed 200 miles away from my home when other pilots (myself included) could have made the landing. It was inconvenient for passengers and crew, and costly in money and public relations for the airline, but it was a classic case of sacrificing lesser goals in the name of safety, and I laud them for it.

Personal minimums only work, however, if you are as strict with yourself as that airline crew was with our flight. It's tempting to try the approach again if others are making it, but you've established personal limits for a reason, and violating your own minimums is putting some other goal ahead of safety.

Your selected route should include the best means of seeing and avoiding aviation weather hazards. For instance, if thunderstorms are forecast, pick a route and an altitude that will keep you in clear air for the greatest part of the journey. You might not get the best ground speed if you fly a bit higher, but that might keep you above the general cloud layer, where you can pick out the thunderheads a hundred miles away and maneuver around them. This is especially important if you don't have airborne radar or electrical discharge detection equipment; you can't always be certain that ATC can see the storms on radar, and you might not have the luxury of leaving frequency to speak with Flight Service.

You'll want to fly high enough to ensure good radio coverage. Flight Watch, properly known as the Enroute Flight Advisory System, is designed to pick up airplanes only above 5000 feet AGL, within 80 nautical miles of remote receiving antennas. Fly lower, and you might not be able to get help from the ground if things go sour. If safety or some other goal dictates a lower altitude, plan to use other FSS frequencies, such as those listening and receiving over VORs, if you need more information.

Make sure you'll have enough fuel for the flight. Bank a little extra for an unexpected headwind or diversions around ice or thunderstorms that add significantly to your time en route. Keep an eye on the clock and your fuel consumption, and divert or land early if you feel your fuel reserve is dwindling. I flew a Bellanca SuperViking from Oshkosh to Wichita one clear October night. According to the winds forecast, I could make it nonstop in about four hours, about maximum for me personally, and I would still have around an hour and a half's fuel in the tanks on arrival. As I flew southward, I encountered mounting headwinds that forced me to recalculate fuel burn and time to destination. Helping me was a Loran; I remember the ETE reading 55 minutes for 5 minutes straight as I plowed into increasing winds over the Missouri/Kansas border. I still have the lapboard sheet I used to refigure fuel remaining at destination; I calculated it at least a dozen times; as the Loran ground speed dropped from a 150 knot still-air reading to under 60 knots across the dark, airportless prairie, I pondered turning around and using the tailwind to return to Kansas City. I even reduced power to 55 percent to increase my endurance, although that slowed my ground speed further. In the end I still had an hour's worth of fuel left when I gassed up at home, but afterward I thought how easily I could have avoided all that stress with a simple fuel stop on the way, and how that added stress and workload could have contributed to an accident if the weather had been worse or for some reason I had to divert over the plains of Kansas.

Another goal is comfort for you and especially your passengers. I see a lot of high-performance aircraft with air conditioners, although using these comfort-increasing devices reduces airplane performance. The *Beech Bonanza* handbook, for instance, says using the air conditioning reduces all performance figures by 10 percent, a significant effect on time and fuel considerations. Here's a case where time and cost are sacrificed for passenger and pilot comfort, but my feeling is that I'd take the extra 10 percent time en route if I could arrive much less fatigued and with passengers happy to fly with me again. I might leave the air conditioning turned off in the climb out of a hot mountain airstrip, however, sacrificing comfort for the safety of improved climb performance in a high-density altitude.

And that's the point. You fly an airplane for many reasons. Foremost in your planning has to be safety, for it does you no good to get over Minneapolis in time for your meeting, only to crash short of the runway because of ice on the airframe. Look at the weather information you receive, as well as factors of airplane and pilot capability, and pick a route and altitude designed to minimize the risks to your flight.

Anticipating weather en route

Once you get your required weather briefing, what do you do with the information? The first part, deciding whether you can go at all, is relatively easy. There are certain weather hazards that spell no-go regardless of the type of airplane you're flying. Extensive areas of thunderstorms, widespread severe or greater turbulence, ceilings and visibilities below legal or personal minimums, freezing rain or snow pellets, and actual moderate or greater rates of ice accumulation are define no-gos for everyone. The presence of aviation weather hazards to lesser degrees might require a flight cancellation, depending on the experience and proficiency of the pilot and/or the equipment on board the airplane. I have different weather minimums, for example, in a Cessna 182 than I do in a Beech Baron, not so much because of the airplane itself but because I'm much more current in the Beech product.

Where the pilot's weather decision-making often breaks down, however, is once he or she makes the initial decision in favor of "go." There are few times when this initial decision can be left to stand without question for an entire flight, yet many pilots do just that: take a weather briefing, by nature the product of experts and therefore seemingly unquestionable, and then launch into the murk without another thought about what lies ahead. The fallacy here is that weather forecasts are etched in stone. Weather is a dynamic, changing process that we don't yet fully understand; forecasts, then, are like works of science fiction. Sometimes they come true but, despite the fact that they're produced with the best information available by some of the best minds in science, sometimes they do not.

You need to treat the weather information you receive with healthy skepticism. There are two types of weather products the briefers tell you about: observations and forecasts. Observations are those that point out the actual conditions that existed at some time in history. The more recent the observation, the better, but any observation is better than a forecast. Forecasts are projections of what possibly might happen in the future. Most forecasting is done by computers, which lump together all the observations, compare the pattern to the last time or times similar patterns existed, and then issue a forecast based on what happened the last time conditions were similar. This is a vast generalization of how forecasts are made, and humans often intervene to add subjective judgment in modifying a forecast, but, again, a forecast is science fiction. You need to take it not as a definite statement of conditions you'll encounter, but as a model of what you'll expect to see, and then compare that model to actual indications en route to verify or refute the forecast.

For example, let's say you're flying from Columbus, Ohio, to our nation's capital. A check of DUAT warns of a cold front extending from around Buffalo, New York, southward across your route, but the forecast predicts only widespread thunderstorms east of the Ohio River. You evaluate your goals and decide to go. You expect to see cumulus-type clouds of extensive development as you near eastern Ohio, and know if things get too bad all you have to do is land short of the line or turn south, away from the low, to look for a path through.

Coming up on the Ohio River, however, not a cloud is in sight. The ride is still smooth; you notice that the tailwind is a little stronger than forecast, and the temperature aloft is a little cooler than the DUAT printout. You conclude that the high-pressure area behind the cold front is pushing eastward faster than expected. What does that mean? Faster fronts usually mean more extensive, and more severe, weather hazards. Turbulence and thunderstorms associated with this front are likely to be stronger than forecast; what's more, the line of activity that was supposed to die out west of Washington might now create hazards on your arrival at destination. It's not time to divert yet, but it is time to talk to Flight Watch to see what's happening ahead.

Let's look at another example. Flying out of Orlando after a family vacation, you expect to encounter a warm front near Atlanta on your way north. Warm fronts mean stratus clouds and poor visibility, but the ride should be smooth. As you fly into central Georgia, you encounter unexpected light turbulence, with an occasional moderate bounce into the headliner. The sky ahead is dark; the south wind is stronger than expected, and the temperature on Macon's ATIS is higher than forecast. The warm front is travelling faster than expected; the air is less stable, and embedded thunderstorms and mixed icing are possible in the clouds ahead. It's time to check for some pilot reports.

When evaluating the weather briefing you receive, then, get out a map and plot the approximate locations of the weather hazards you expect to see. Note the types of clouds you expect because clouds are the visible signposts of weather hazards. To have for comparison later, jot down the forecast temperatures at surface-reporting points and aloft for your route. En route, if you see more or less extensive cloudiness, or clouds of an unexpected height or type, you can determine if a front is moving faster or slower than expected, and you can anticipate a change in weather hazards. Hotter temperatures might mean more moisture in the air; depending on your direction relative to the air masses, a faster warm front or slower cold front indicate increased flying risk. Colder-than-expected temperatures aloft or at the surface might mean ice where none was forecast, or the colder temperatures might mean a faster-moving cold front, which usually means more severe weather. Winds other than expected might mean more or less moisture for condensation or might indicate the speed of any fronts, so anticipate your ground speed and compare it to what you really get. Different winds also mean refiguring fuel consumption, of course.

While you're preparing for the flight, go ahead and highlight frequencies for Flight Service and ATIS or automatic weather stations you'll overfly. You'll need to get the AWOS/ASOS frequencies from the Airport Facility Directory, instrument approach charts, or commercial sources; it'll be a lot easier to do this on the ground instead of shuffling paperwork in flight. If your flight time overlaps the issuance of area forecasts or the convective outlook, for instance, note the point in your flight when you'll call for an update. Arrange all this information where you can easily reference it in flight; you'd hate to do all this work, only to have to look it up again in a dark, turbulent cockpit.

Just before takeoff

When you're about ready to go, call Flight Service or reference some other approved weather source for an update. You really want two bits of information: current conditions and pilot reports. Compare the current conditions to what your earlier briefing's forecast called for. Are things going as expected, or is there a significant difference? If things aren't working out as expected, ask the briefer why. If you can't get a good explanation, ask yourself what might cause visibility to lower, for instance, or thunderstorms to form away from a front. You can use your knowledge of dry lines or upslope conditions to your advantage this day.

Pilot reports are the only accurate means of locating turbulence, icing, and cloud tops, so the more of reports you can get, the better off you'll be. Make a vow to issue at least one pilot report per hour en route, regardless of conditions. Wouldn't you like to hear a PIREP of "negative icing" at least as much as one calling for light icing in the clouds?

Based on the abbreviated briefing you receive, revise your expectations of the weather phenomena you'll see and, as necessary, plan a change in route and/or altitude to minimize the risks involved. Recheck your fuel calculations with updated winds in mind, and file an amendment to your flight plan if conditions warrant.

En route

Once airborne, your job is easy. You've already done a lot of planning before takeoff, so from a weather standpoint, all you have to do is to look outside and see if things are happening as expected. Is your ground speed close to predictions? Are you seeing the anticipated pattern of clouds? Do temperatures aloft and at reporting points along the way agree with the forecast? Modify your expectations occasionally, reroute as necessary, and always ask yourself how each change will affect your goals for the flight.

File pilot reports. I like to "trade" information with Flight Watch or Flight Service when I call in. "This is Bonanza 12345 with a pilot report; advise when ready to copy," is the first contact I usually make. Flight Service folks are extremely happy to hear from pilots, so by giving them a PIREP first, you'll "spring-load" the briefer to give you help later. Report anything that you'd want to know about conditions if you were calling in for an initial briefing. Address the four hazards: thunderstorm activity in sight or on radar or your Stormscope, turbulence you've encountered along the way, your best estimate of flight visibility as well as cloud bases and tops, and, of course, the presence or lack of any airframe ice. You might also want to pass along outside air temperatures and computed wind speeds and directions (most Lorans and GPS units can calculate this for you). Don't forget to report your location and airplane type. Give Flight Service anything that will help the next pilot make an informed decision. Then ask for the updated information you want.

Pilot reports made to Center controllers might or might not make it into the Flight Service briefing network, so to do the most good, file PIREPS with Flight Watch or Flight Service. Approach and tower controllers have no mandate at all to pass on pilot reports, so, although it's a good idea to mention your data to these controllers for the immediate benefit of others flying in the area, take a moment as time permits and call the information in to Flight Service. Of course, on landing you might encounter wind shear or some other hazard you want to report, but you don't have time to call it in. As an experiment once I called Flight Service on the phone after landing to provide a "postdated" pilot report, warning of light ice on final approach. I gave the time I actually encountered the ice, instead of the time of my call, because that's what was valid; the briefer took the information happily, as if she did this every day. The only possible problem with this method of getting safety information to pilots is the long wait you might experience waiting for a briefer; I suppose you could give the PIREP on the "fast-file" tape recording. That's obviously not the intended use of the recording, but I imagine the Flight Service people, who almost beg for pilot reports, wouldn't mind.

15
Sample flights

LET'S RUN THROUGH AN EXAMPLE OF HOW I OBTAINED AND EVALUATED weather information for that trip from Wichita to Minneapolis.

TRIP ONE: WICHITA TO MINNEAPOLIS

I had at my disposal an F33A Bonanza, well equipped with IFR avionics, a Stormscope, and an IFR-certified (for en route, anyway) GPS receiver.

I knew about three days in advance that I would make the flight Friday afternoon, weather permitting. This gave me the chance to keep an eye on weather forecasts from several sources to compare them against one another and themselves. Mass-dissemination weather outlets (newspapers, local and cable television, National Weather Service radio, etc.) get their information from different sources. Look in the margins of a printed forecast, or listen to the weather person to determine if the forecast is his or her own or if it comes directly from the National Weather Service. Everyone has exactly the same information, the official observations, to start with. If there is a difference in one source's forecast as compared to another's, it's because no one is really certain what will happen. If all the sources give about the same weather scenario, it's much more likely to come true, but if there are big differences, assume the unexpected. As the days go on, if a single source changes its forecast, then their computer model might be in doubt. The Weather Channel, for instance, does its own forecasting for whenever a person stands in front of a map, but you'll notice the narrator of the local forecasts mention that these come from the National Weather Service. That's why there's often a discrepancy between the two.

Figure 15-1 shows the Wednesday, Thursday, and Friday newspaper forecasts for weather in the Wichita area, as well as a general map of the national weather. The forecast for Friday was quite consistent; in fact, the original forecast for thunderstorms was downgraded to simply clouds and wind. I did wonder a little about the stationary front that "appeared" Thursday and "disappeared" Friday along my route of flight.

The TV news crews, both locally and on cable, called for a similar forecast. Since everyone seemed to be in general agreement, I felt pretty good about the flight, expecting cloudiness and wind along the way, with some precipitation and the possibility of a storm near my destination. Because it was winter, I was concerned about airframe ice, but the weather was unseasonably warm, so my fears were limited to the northern half of my journey.

Thursday afternoon I called Flight Service for my first "official" look at the weather. When calling Flight Service, give the briefer all the information he or she needs to serve you best. This is what they expect to hear:

- Your pilot qualifications (IFR or VFR).
- Whether the flight will be VFR or IFR.
- Aircraft type and registration number.
- Departure point, proposed route and destination.
- Proposed altitude.
- Estimated time of departure.
- Estimated time en route.
- Type of briefing requested.

I've called Flight Service in several parts of the country where the briefer won't start unless you've already filed a complete flight plan. That's backwards in my thinking because you can't figure an accurate time en route, altitude, route of flight or, in some cases, true airspeed, unless you already have the weather data. I've found the path of least resistance is to have a general flight plan in mind before you call. You can give that plan if the briefer insists, and update it later once you have the information you need.

So I called Thursday afternoon and checked in: "I'm an IFR-rated pilot planning an IFR flight in a Bonanza 12345. I'm leaving Wichita, flying directly to Minneapolis,

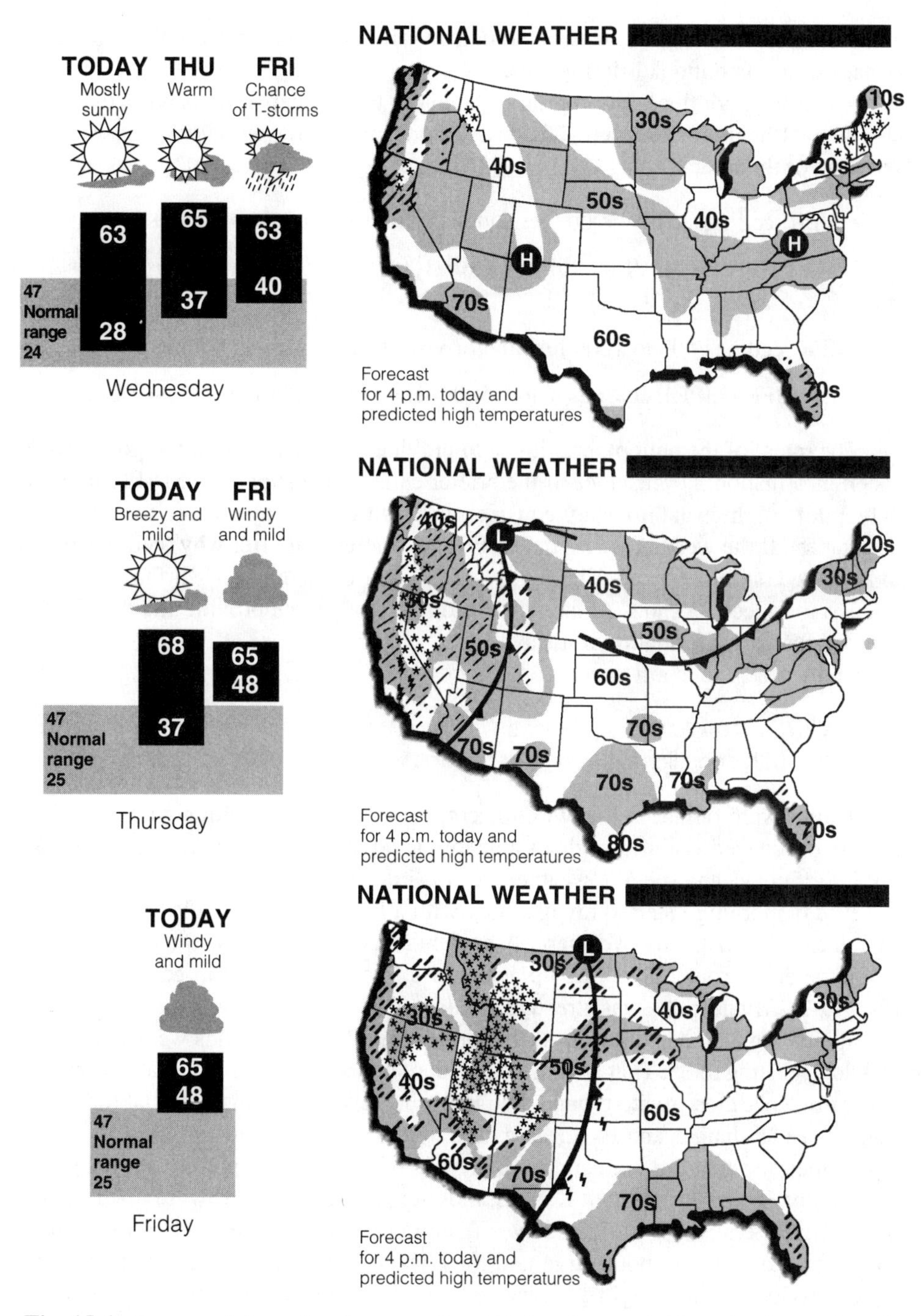

Fig. 15-1. *Commercial weather forecasts from Wednesday (top), Thursday (center), and Friday (bottom) prior to a Friday flight.* WeatherData, Inc., in the Wichita Eagle

altitude unknown for now, leaving tomorrow around 1800 zulu for three hours en route. I'd like an outlook briefing please."

An outlook briefing is appropriate when your time of departure is more than 6 but less than 24 hours in the future. The briefer will give you essentially a one-word description of the forecast:

- VFR: ceilings above 3000 feet and visibilities greater than 5 miles.
- Marginal VFR (MVFR): ceilings from 1000 to 3000 feet and/or visibilities from 3 to 5 miles.
- IFR: ceilings 500 to 1000 feet and/or visibilities from 1 to 3 miles.
- Low IFR (LIFR): ceilings below 500 feet and/or visibilities below 1 mile.

The intent of the outlook briefing is to provide pilots with the initial go/no-go decision information. For instance, if the briefer called for IFR or even MVFR, a visual-only pilot might want to start considering a different way to get to his or her destination. If the weather is less than VFR, the briefer will offer why (ceilings or visibility) the conditions exist. The word "wind" will be included if surface winds are forecast to exceed 25 knots. At their discretion, briefers might provide additional, more detailed forecast information if it's available.

In my case, the report was for:

"MVFR in the Wichita to Kansas City area due to ceilings, wind, dropping to MVFR with occasional IFR due to ceilings in northern Iowa and Minnesota."

Things were falling in line with the commercial forecasts, although, with the exception of reduced visibility and possible low-level, wind-induced turbulence on takeoff, I didn't have any specific weather hazards information yet.

Friday morning I started my planning with a look at the cable TV broadcast. At 9 a.m. local (1500Z), The Weather Channel indicated a strong, southerly wind flow along the entire route, air piling up ahead of a strong cold front in central Colorado (Fig. 15-2). Minneapolis was already near its high temperature for the day, in the mid-40s, indicating that the cold front was approaching. Wichita's winds were southerly at 26 knots, gusting to 35. That sounds bad, and, of course, calls for moderate or even severe turbulence close to the ground, but, unfortunately, it's not that unusual for Kansas as the seasons change, and I've grown used to it.

Something that did concern me a bit was the local radar picture showing scattered, light echoes throughout Kansas, moving predictably northward at a high rate of speed. Coverage was indicated over Wichita itself despite the lack of rain falling, so I began to wonder about virga and the chance of a microburst on takeoff. Scattered activity like that might also warn of cumulus clouds, indicating a greater likelihood of icing.

Fig. 15-2. *The general location of weather influences relative to the proposed route of flight.*

I called the ATIS recording at Wichita Mid-Continent Airport (the phone number is in the Airport/Facilities Directory). It reported:

"1456 zulu: temperature 54, dew point 41, wind 170 at 22, altimeter 29.83."

Since I know that Wichita's ATIS comes from an AWOS, or automated source, I know that the absence of ceiling and visibility information doesn't mean that the weather's clear. It just denotes "no clouds below 12,000 feet, visibility greater than 10 miles." I called the AWOS at my departure airport, Colonel James Jabara, on the east side of the city about 10 miles from Mid-Continent. The Jabara AWOS, which at the time was still in the testing stage, called for:

"1508 zulu: sky clear below 12,000, visibility greater than 10 miles. Temperature 55, dew point 40, wind 160 at 23 peak gust 28, altimeter 29.80."

There was the cause of the wind, I thought, a three-hundredths of an inch drop in pressure over only 10 miles. The wind, and any associated turbulence, wouldn't go away soon.

I haven't invested in a modem for my home computer yet, so I can't use DUAT. Instead, I rely on a call to Flight Service for my weather information. When I'm not in a real hurry, as was the case this day, I like to drive over to Mid-Continent Airport and visit the FSS personally. I respond well to visual displays, and like to ask the briefers face-to-face questions about their charts. Along the way I took a critical look at the clouds. There was a mid-altitude layer of stratus, scattered to broken, racing north with the wind. Above that was a high-altitude cirrostratus layer. The sky was dark to the north, along my route. As expected, there were a few cumulus bumps rising out of the stratus. This might be a challenging flight.

Arriving at the Flight Service Station (the pilot briefing areas always seem to be deserted), I gave the requisite information and asked for a Standard briefing. This is what the briefers are trained to provide in a standard briefing:

- Adverse conditions: any hazardous reported or forecast conditions, such as AIRMETs or SIGMETs, or other hazards advisories.
- The statement "VFR Flight Not Recommended," if the pilot requests a VFR briefing and conditions are currently or forecast to be below VFR limits.
- Synopsis: the type and location of fronts, pressure systems, and other weather producers.
- Current conditions: unless the pilot's estimated time of departure is more than two hours in the future, the briefer will provide current conditions at the departure and destination airports, as well as any en route the briefer deems pertinent.
- Forecast conditions: for those same reporting points.
- Winds aloft.
- Notices to airmen: outages, closures, or changes in procedure or regulation that might affect the flight.
- Known ATC delays.
- Additional information: special-use airspace warnings, as well as any other information the pilot requests.

The intent of the standard briefing is to provide the pilot with all the weather information he or she needs to safely plan a flight. I've condensed the reams of paper I received from Flight Service into the pertinent reports and my comments:

Flight precautions

"SIGMET NOVEMBER 1 valid until 182200 SD NE KS IA MO OK
Occasional severe turbulence below 5000 AGL due to strong southerly wind flow."

I expected that.

"CONVECTIVE SIGMET - None. No thunderstorms expected, valid through 2255."

Good news, somewhat unexpected because of the fast-moving front.

"CENTER WEATHER ADVISORY for the Kansas City Center region, valid until 1900Z, for frequent moderate and isolated severe turbulence from the surface to 3000 feet."

Again, not unexpected.

> "AIRMET TANGO for turbulence: low-level wind shear KS NE MN IA until 2000Z. Occasional moderate, isolated severe turbulence below 6000 feet due to low-level winds, to continue beyond 20Z."

This was nothing new to me. Although it didn't directly affect my flight, it also included the following:

> "Occasional moderate turbulence 18,000–39,000 feet due to jet stream and upper level trough."

There's where my cumulus were coming from! The jet stream was very high for winter, however, and wouldn't create a turbulence hazard for my Bonanza.

> "AIRMET ZULU for icing and freezing level: NE KS: occasional moderate rime or mixed icing in clouds and precipitation between 10,000–12,000 and 18,000 feet. Conditions developing and spreading northward during the period. Elsewhere, no significant icing outside of convective activity. Freezing level 8000–10,000 northeast of Des Moines and sloping to 10,000–12,000 south of that line."

It looked good. I could climb through the low-level turbulence. My cruise climb speed approximates turbulent air penetration speed, and I can fly below 10,000 most of the way, above the turbulence but below the ice. If I get into clouds north of central Iowa, I'll descend to 7000 feet. That should get me below the freezing level but still keep me above the worst of the turbulence.

Synopsis

> "A Quasi—stationary front extends from the northern Great Lakes region across southern Manitoba and northern Montana. Pacific frontal system over the Rockies will move eastward during the period."

This stationary front north of my route might serve to trap moisture pushing up from the south along the cold front, which could be the reason for the forecast reduced visibilities in Minnesota. In pictorial form the system, and my proposed route, looked like Fig. 15-2.

These were the area forecasts:

> "KS AGL 30-50 SCT 100-120 SCT-BKN 200. CI ABV. SLY WINDS 15-20G25-35. OTLK MVFR CIGS/RW/TRW/WIND."
> "CNTRL AND ERN NEB AGL 30 SCT 100-120 SCT-BKN 200 CI ABV. SLY WINDS 15-20G25-30. OTLK VFR."

"IA OTLK MVFR CIGC RW."
"MN 17Z-23Z BCMG AGL 5 SCT CIGS 10-20 BKN-OVC. OCNL R-L-F.
OTLK IFR CIGS R F."

The Kansas area forecast still called for thunderstorms, which I'd have to get more information about; the rest of the route looked okay except for lowering visibilities on the destination end. I had an altitude in mind already, to avoid turbulence and icing. I needed to work on a direction to improving conditions.

Surface aviation reports

"ICT SA 1556 A02A M43 BKN 10+ 084/56/41/1924G32/979/ PK WND 1832/1531."

Mid-Continent was reporting a 4300-foot broken layer, significantly lower than the ATIS an hour before but making the forecast MVFR ceiling more likely. Visibilities were good, and the wind was still strong and gusty from the south. Air pressure continued to fall. The front was approaching steadily.

"3KM SA 1550 E100BKN 10 60/M/1820/980."

Colonel James Jabara Airport confirmed the wind and pressure drop. Clouds were higher on the east side of town.

Along the route, in central and northern Kansas, I had:

"EWK (Newton, KS) 1556 AWOS 39 SCT 10 54/37/1827G36/979" "SLN (Salina, KS) 1550 E120OVC 7 054/55/40/1825G31/970."
"MHK (Manhattan, KS) 1550 E110 BKN 250 OVC 7 074/59/41/2018G35/975 PK WND 2040/20."

The low clouds seemed to be a local phenomenon, as Newton is only a few miles north of Wichita. The wind was consistently gusty out of the south. Through Kansas, then, I could expect primarily a mid-level overcast, with maybe a few cumulus thrown in, and low-level turbulence.

Farther north, I had more information:

"OMA (Omaha): 1550 120 SCT E150 OVC 7 063/54/40/1724G31/971."

More of the same over Nebraska, with a lowering visibility.

"FOD (Fort Dodge, Iowa): 1556 AWOS CLR BLO 120 10 45/37/1817G22/974."
"MSP (Minneapolis): 1552 120 SCT 250 -OVC 7 073/46/31/1715G22/973/HAZY."

Except for the wind, both were still better than the area forecast called for.

Terminal forecasts

"ICT 16Z 30 SCT C70 OVC 1928G38 OCNL C30 OVC CHC 4RW-/TRW-AFT 20Z."

Thunderstorms, undoubtedly brought on by the trough aloft, were still considered possible, but not for two hours after my planned departure.

"MHK (Manhattan, KS) UNTL 20Z C120 BKN 2020G30."

More of the same.

"OMA (Omaha) 18Z 40 SCT 120 SCT 250 -BKN 1820G30."

Not bad.

"MSP (Minneapolis) 20Z 4 SCT C8 OVC 2F 1615G25 OCNL C4 BKN 1R-L-F."

This might be a bit of a challenge: IFR weather in light rain, light drizzle, and fog. I need an alternate. I know weather will remain good to the south and east, so I'll look for a field in southern Wisconsin. It looks like Madison should be "C60 BKN 1614 CHC RW- until 02Z6," so that's my alternate, at least for now. I can always change that if I need to in flight, but I'll plan to have enough fuel to make it to Madison if I need to miss the approach at Minneapolis.

Winds aloft

I planned on filing for 9000 feet to meet weather requirements and the hemisphere rule, but I asked for the winds at 6000 and 12,000 feet as well so I'd have a better idea of what was above and below me. Table 15-1 shows what I got.

Table 15-1. Winds aloft

Station	6000	9000	12000
ICT	2146+08	2346+05	2350–02
OMA	2157+08	2252+04	2257–03
MSP	2255+10	2255+03	2155–05

This confirms my decision about altitudes. In fact, I don't have to worry about the freezing level dropping below my 9000-foot cruising altitude for the entire length of the trip, despite the freezing level forecast. I'm glad I'm flying north today!

NOTAMS

There were a few runway closures and an AWOS out of service at fields along my route of flight, which I noted, but nothing significant except that the Minneapolis VOR was to be out of service. I looked at the approach plates for my destination airport, Flying Cloud Airport on the southwest side of Minneapolis, and decided that the Minneapolis VOR outage wouldn't affect me.

There are three types of NOTAM, and two classes of NOTAM (Fig. 15-3). The three types are:

```
******** NOTAMS ********
!ICT 02/064 ICT 1L-19R CLSD
!ICT 09/004 ICT 1L ALS UNMON
!FRI 02/002 FRI ATCT HRS TIL 2200
!FDC 4/0641 Y12 FI/T LAKEVILLE/AIRLAKE, LAKEVILLE, MN..
ILS RWY 29 AMDT 2...
S-ILS 29 VIS 1 ALL CATS.
S-LOC 29 VIS CATS A/B/C 1.
FOR INOPERATIVE MALSR INCREASE S-LOC 29 CAT C VIS TO 1 1/4 MI
AND CAT D VIS TO 1 1/2 MI.
OBTAIN LOCAL ALSTG FROM AWOS (115.7).
MISSED APCH: CMB TO 1500 THEN CMBG LT TO 2800 DRCT FGT VORTAC
AND HOLD.
VOR-A AMDT 3...
MISSED APCH: CMB TO 2700 DRCT FGT VORTAC AND HOLD E, LT, 265
IBND.
OBTAIN LOCAL ALSTG FROM AWOS (115.7).
```

Fig. 15-3. *Samples of NOTAMs.*

NOTAM L. Local NOTAMs, dealing with relatively minor outages and closures not likely to affect your decision to make a flight. This might include VFR approach aids (VASI, PAPI, etc.) out of service, taxiways closed, and the like. I heard a local NOTAM on ATIS once that the airport restaurant at Muncie, Indiana, was closed for the day. Local NOTAMs cue up automatically during briefings only from the "owning" Flight Service Station. In other words, a local NOTAM for an airport in the Indianapolis Flight Service Station area would be reported during a briefing to anyone who called the Indianapolis office, but not if the pilot was speaking with another Flight Service outlet. You can, however, ask your briefer to call up any local NOTAMs for areas outside their own jurisdiction; they're available for the asking.

NOTAM D. Distant NOTAMs will automatically be reported during a briefing regardless of areas of jurisdiction. These NOTAMs are issued for any situation (runway lights out of service, airport closed for snow removal, etc.) that might affect your decision to fly into that airport.

FDC NOTAMs. The Flight Data Center is that agency charged with creating and maintaining visual and instrument en route and approach procedures. If a change occurs (an MEA or MDA change, perhaps, or a navaid frequency update), an FDC NOTAM will be issued until the next round of charts, bearing the updated information, is published. FDC NOTAMs are disseminated everywhere, like NOTAM D's.

The two classes of NOTAMs are:

Class I NOTAMs. NOTAMs issued for short-term situations, usually expected to last less than two weeks. Class I NOTAMs L, D, and FDC will be briefed as previously indicated when you obtain weather data.

Class II NOTAMs. NOTAMs lasting longer than two weeks are published in a book distributed to all Flight Service and ATC facilities. Flight Service briefers will assume you've read the Class II (sometimes called the "published") NOTAMs unless you specifically tell them you have not. A VOR might have been decommissioned, then, or an airport closed, and you won't hear about it if it's in the Class II book. This is a throwback to the days when there were weather outlets just about everywhere, and most weather briefings were conducted face-to-face. Unless you've read the book in the last few days, be sure to ask for the Class II NOTAMs when you get your briefing.

I still had a few unanswered questions, so I used my knowledge of aviation weather and queried the briefer further. Because the possibility of thunderstorms in Kansas cropped up in a couple of places, I asked for the stability indices for my route of flight. The air ahead of the front was extremely stable, I found. Along my route, the lowest lifted index number was +9, and the most prevalent was +11, so any cumulus activity was the result of low-level turbulence and was not likely to develop significantly. I asked for the convective outlook, just to be sure, and it echoed the convective SIGMET statement that no thunderstorm activity was forecast.

I was still a little wary of icing, so I asked for the 700-millibar constant pressure chart. At Flight Service they even color-code this chart, which approximates the 10,000 foot level, near my altitude of choice. If the temperature/dew-point spread is within five degrees, the chart is colored green. According to the chart displayed, there was little chance of clouds at 9000 feet anywhere along my route of flight. In fact, except for a few mid-level clouds on departure and the possibility of low clouds at my destination, I could easily make this trip VFR.

Finally, I wanted more information about the jet stream, just in case it might be low up north or would add to lifting action to create a storm. The tropopause height/vertical wind shear chart wasn't currently available, said the briefer, but I found that I could get the same information by asking for the 300-mb, or 30,000-foot, constant pressure chart. It clearly showed that no jet stream cores had formed any closer than California. In reviewing this jumble of information, then, I evaluated the four aviation weather hazards:

Thunderstorms: extremely unlikely. Any that might form were likely to be small and isolated. If I flew at 9000 feet, which should keep me in clear air, I could visually avoid any build-ups. My Stormscope would be a help, but I expected it to be blank all the way to Flying Cloud.

Turbulence: extremely likely down low. For takeoff, anticipating wind shear and even the possibility of a virga-inspired microburst, I'd use all the available runway and my best short-field technique, rotate to and climb at my Vx pitch attitude until reaching 1000 feet above ground level, and be wary of sudden airspeed gains or losses on climbout. I don't normally use flaps on takeoff but, if I did, I wouldn't retract them until reaching 1000 AGL either. After reaching 1000 AGL, I'd stay at full power and Vy airspeed and attitude, getting me through the turbulence layer below the airplane's turbulent air penetration speed until I reached smooth air. At destination, I'd ask to remain high as long as possible, and slow to approach speed prior to descent. This would keep me below turbulent air penetration speed and minimize my time in the bumps. To increase stability, I might also extend the landing gear for the descent.

Reduced visibility: shouldn't be a factor until destination; I already had an alternate in mind, with plenty of fuel to get me there.

Ice: not likely, as my cruising altitude keeps me below the freezing level and, for the most part, out of the clouds. I'll watch for cumulus clouds along the way and try to avoid them. I'll also watch for ice formation if I fly into any rain. Temperatures down low are warm, as much as 10 above Celsius at 6000 feet over Minneapolis, so if I ice up I'll slow to turbulent air penetration speed and ride out the bumps down low until the ice melts. If that doesn't work, any turn towards the east should get me out of the rain and into warmer air.

I thought I had all the bases covered. I filed a flight plan GPS direct to Manhattan, Kansas (avoiding a corner of the Ada East MOA, in north central Kansas), then direct to Flying Cloud airport (Fig. 15-4).

Form Approved: OMB No. 2120-0026

U.S. DEPARTMENT OF TRANSPORTATION FEDERAL AVIATION ADMINISTRATION FLIGHT PLAN	(FAA USE ONLY) ☐ PILOT BEARING ☐ VNR ☐ STOPOVER	TIME STARTED	SPECIALIST INITIALS

1. TYPE	2. AIRCRAFT IDENTIFICATION	3. AIRCRAFT TYPE/ SPECIAL EQUIPMENT	4. TRUE AIRSPEED	5. DEPARTURE POINT	6. DEPARTURE TIME PROPOSED (Z)	ACTUAL (Z)	7. CRUISING ALTITUDE
VFR / X IFR / DVFR	N12345	BE33/R	149 KTS	3KM	1800		9000

8. ROUTE OF FLIGHT

3KM direct MHK direct FCM

9. DESTINATION (Name of airport and city)	10. EST. TIME ENROUTE HOURS	MINUTES	11. REMARKS
FCM	2	15	IFR enroute GPS equipped

12. FUEL ON BOARD HOURS	MINUTES	13. ALTERNATE AIRPORT(S)	14. PILOT'S NAME, ADDRESS & TELEPHONE NUMBER & AIRCRAFT HOME BASE	15. NUMBER ABOARD
5	00	Madison, WI	Thomas Turner on file at ICT	1
			17. DESTINATION CONTACT/TELEPHONE (OPTIONAL)	

16. COLOR OF AIRCRAFT	
white/blue	CIVIL AIRCRAFT PILOTS. FAR Part 91 requires you file an IFR flight plan to operate under instrument flight rules in controlled airspace. Failure to file could result in a civil penalty not to exceed $1,000 for each violation (Section 901 of the Federal Aviation Act of 1958, as amended). Filing of a VFR flight plan is recommended as a good operating practice. See also Part 99 for requirements concerning DVFR flight plans.

FAA Form 7233-1 (8-82) CLOSE VFR FLIGHT PLAN WITH__________ FSS ON ARRIVAL

Fig. 15-4. *Flight plan filed from Wichita Colonel James Jabara airport to Minneapolis' Flying Cloud airport.*

Ready to go

Notice the true air speed on my flight plan. Doesn't 149 knots seem a little slow for a Bonanza? That, too, is a result of my goal-setting process.

Safety is my overriding goal, to be sure, but beyond that other goals are interchangeable. One priority is time. With a whopping tailwind like I expect, I'll make very good time to Minnesota. So good, in fact, that I can reduce power and still make it in short order. Having chosen 9000 feet and a route based on weather considerations, I took out the *Pilot's Operating Handbook* for the F33A and did a little comparison of the effect of using different power settings. Table 15-2 shows what I came up with.

Table 15-2.
The effect of different power settings

Power setting	TAS	Time	Fuel burned
75%	169 Kts	2:03	29 Gal
65%	161 Kts	2:08	27 Gal
55%	149 Kts	2:15	25 Gal

By using a lower power setting (full throttle, 2100 rpm), I take only 7 minutes more to destination over the "traditional" 2300 rpm, 65 percent setting, with a 2 gallon, or 8 percent savings in fuel. In fact, I'm airborne only 12 minutes, barely perceptible, more than over the highest power setting, and I save 14 percent in fuel. The lower engine rpm significantly reduces noise in the cockpit, heightening the goals of comfort and enjoyment. Most engine experts agree that the lower the rpm, the less wear an engine experiences, so cost goals are met in both direct fuel burn and long-term engine maintenance. Hence the lower power setting for this flight, meeting many goals at once without detracting from safety.

Just before takeoff, I called Flight Service once more and asked for an abbreviated briefing. An abbreviated briefing, designed to fill in the latest updates and answer any other questions after a standard briefing from any other source, includes a review of any flight hazards, current conditions for departure, en route and destination reporting points, pilot reports, and any questions the pilot might have. My abbreviated briefing included the AIRMETs and SIGMETs I've already related, and the following:

Current conditions

"ICT 1655 M41 OVC 10+ 58/38/1830G36/975."

Stronger winds, a change to overcast skies, and a continuation of the big pressure drop.

"MHK (Manhattan, KS) 1650 E110 BKN 250 OVC 5BD 062/61/40/2018G39/972 /PF WND 1942/32."

Strong surface winds, combined with the dry weather we'd had lately, led to reduced visibility (but still five miles) in blowing dust.

"OMA E140 BKN 7 047/56/40/1722G28/967."

More of the same.

"FOD (Fort Dodge, IA) CLR BLO 120 10 48/39/1818G28/971."

"MSP 120 SCT 250 -OVC 7 066/47/31/1715G20/971/HAZY."

With the exception of the overcast sky at Wichita, then, the forecast en route seemed to be holding up, and conditions at Minneapolis were significantly better than the IFR conditions forecast.

Pilot reports

These are what I really like to hear in an abbreviated briefing because pilot reports are the only true indicators of what is happening aloft. I've edited and commented on the PIREPs.

Over ICT: "FL030/TP LEAR/WV 176049/FL040 WV 196069/FL06020060." Winds were even stronger at altitude than expected. That means even more intense turbulence, possibly, and a quicker trip to destination.

Over ICT: "FLKN/TP BE02/SK 053 BKN 055." A King Air reported the bases at 5300 feet, which confirmed the Wichita ATIS report of around 4000 above ground level. It was a thin, broken layer only 200 feet thick, with no mention of ice. Good news.

Over ICT: "UUA (Urgent pilot report, for severe conditions): FL030/TP C172/TB MDT-SVR 2 NE ICT ARPT." A local Skyhawk was being battered on arrival at Mid-Continent.

"FOD 1635 FLUNKN/TP BE58/TB MDT BLO 1000 AGL." A Baron at Fort Dodge confirmed the low-level turbulence there.

"FOD 1612/FL035/TP C172/WX FV10H/TA 56F/TB SMTH." Okay! The ride was smooth as low as 3500 feet over Fort Dodge. Visibility was 10 miles and the temperature was almost tropical for Iowa in winter.

"FOD 1603/FLUNKN/TP C560/WV 1620 MSL 183049." A Citation confirmed the strong wind flow.

"MSP 1600/FL020/TP OH58/SK 200 SCT/WX FV7H/TA 08/WV 19030/TB LGT OCNL MDT." A helicopter reported visibilities, cloud cover and turbulence significantly stronger than forecast at my destination.

I also asked for any radar activity, and I was told that there were no echoes anywhere along my route. For that matter, there was no precipitation east of about the Colorado/Kansas line.

My abbreviated briefing, then, confirmed that the turbulence was my greatest threat. Thunderstorms seemed nonexistent. Chances for icing and reduced visibility near the surface also looked to be remote. My initial look at the weather warned of all four weather hazards: thunderstorms, turbulence, reduced visibility, and airframe ice. Now things looked much less dire, and I had the confidence to make the final ground-based decision in favor of "go."

En route

My takeoff roll was ridiculously short into the monstrous wind. Turning north to leave Jabara Airport, I watched the GPS ground speed readout climb well above 200 knots.

As expected, turbulence was fierce, but not unmanageable, and I soon climbed into smooth, stable air above 3000 feet. I had been cleared "as filed" and, with a short delay at 3000 feet for traffic, was given my requested 9000-feet cruising altitude.

The first layer of clouds was about where reported, too, with bases at 5100 and tops of what east of Wichita was a scattered layer around 5400. I made a note of the time and altitudes to call in as a pilot report once I leveled off and settled into cruise. Reaching 9000 feet well north of Wichita, propelled by the southerly wind, I could see the northern limits of the lower cloud layer; I was over Manhattan in no time, turning slightly to fly directly towards Flying Cloud Airport under a high, broken cloud layer.

In only an hour I was approaching Omaha, and I tuned Omaha's AWOS for an update:

"Omaha AWOS 1906 Zulu Clear below 12 thousand, visibility greater than ten miles. Temperature 61, dew point 38, wind 170 at 24 gust 33. Altimeter 29.58."

The temperature was warmer than expected. Winds were about as anticipated, and the drop in air pressure told me that the front was moving faster than expected. Ahead there was only the broken cloud layer above me, although the sky westward was dark. It was after the hour, so I called Flight Service for an update. I provided a pilot report, and I asked for en route weather and pilot reports. I received:

"AIRMET TANGO AMENDMENT 2 FOR TURBULENCE: Occasional moderate, isolated severe turbulence over Nebraska, Iowa and Minnesota below 6000 feet due to increasing low-level wind flow."

This amendment was merely an extension of the earlier AIRMET for this hazard, which was due to expire. All other AIRMETs and SIGMETs remained unchanged.

Surface observations

"Fort Dodge, Iowa: Clear below 12,000, visibility greater than 10 miles, wind 180 at 23 gusting to 33."
"MSP: Estimated ceiling 13,000 broken 25,000 overcast, visibility 7. Temperature 45, dew point 31, wind 170 at 15 gusting to 22, altimeter 29.64."
"Flying Cloud Airport (FCM): Estimated ceiling 15,000 overcast, visibility 8, wind 160 at 16, altimeter 29.60."

Conditions were holding along my route of flight, and they were still much better than forecast over Minneapolis. Perhaps the strong southerly winds, stronger even than anticipated at altitude, were blowing the moisture expected to blanket Minnesota further north, or maybe the winds prevented moisture from the Great Lakes from extending as far west as forecast. I made a mental note that my alternate in Wisconsin might be lower than forecast if the latter were the case; although having to miss the approach at Flying Cloud no longer seemed likely, I might have to head back south, into the gale, to find better weather if a miss were required.

Pilot reports over FCM

"Flight Level and Type unknown: turbulence light to moderate from the surface to 2000 feet."
Over MSP: "Flight Level 7000 feet: a Saab 340 reported the top of the haze layer at 4000 feet, light to moderate turbulence below 4000 feet and smooth above, during a climb northbound."

I was lucky that day. Things were getting better all the time. Flight Service did report another wrinkle, however. Des Moines radar had detected a new set of light rain shower cells on the 248 radial at 87 miles, with maximum tops to 18,000 feet. That was already southeast of me, however, and I didn't see any other indications of precipitation.

In less than half an hour more, I was given a vector to intercept the Meinz 6 arrival into Minneapolis for flow control; this was a small turn to the right near Fort Dodge, Iowa. I listened to weather:

"Fort Dodge Automated Weather 1937 Zulu. Clear below 12 thousand, visibility greater than ten miles. Temperature 51, dew point 42, wind 160 at 17 gusting to 26, altimeter 29.62."

In another ten minutes I could receive Information Uniform on Minneapolis/St. Paul International's ATIS:

"Minneapolis/St. Paul Information Uniform, 1950 Zulu observation. Estimated ceiling 14 thousand overcast, visibility 8. Temperature 45, dew point 31, wind 160 at 17, altimeter 29.61."

There were no surprises. I started setting up for my arrival.

Soon thereafter I was vectored for a visual approach into Flying Cloud, a long downwind to runway 18. I still expected turbulence down low and, not liking to bump around all that much, I was granted my request to stay above 3000 feet until near the south boundary of the airfield. Slowing to approach speed in the still-smooth air, I began my final descent by lowering the landing gear, reducing power slightly, and using this slow speed to come down at a steep angle through the turbulence while remaining quite stable owing to the airplane's speed and configuration. Sure, I rocked around a bit in the last few minutes of flight, but it was much more comfortable than an extended period "down low" with a higher airspeed. My computed ground roll was only 400 feet into the southerly wind.

Hopefully, this example shows how you can obtain weather information and properly use it in making and evaluating your go/no-go decision. There were a lot of variables on this flight, so many as to potentially overwhelm me with data, so I broke it down into the four classes of aviation weather hazard. Then I derived a plan, a route, and an altitude designed to limit my exposure to each of those hazards. I also planned an escape route towards improving weather should any problem arise, so I could react to changing conditions quickly enough to avoid being a statistic.

With the weather information in hand, I also looked at some of my other priorities, or goals, for the flight, and I chose a somewhat unconventional power setting that reduced fatigue and noise level, extended engine life, saved a few gallons of fuel, and still got me to my destination in about the same amount of time as a higher power selection.

Ready to go, I didn't merely take what nature and my initial weather briefing gave me. I called for an update, and I regularly received new information en route to refute the dismal forecast for my arrival. Remember, it could have easily gone the other way. For example, if I found over Iowa that the Minneapolis area was socked in, with ice or thunderstorms hidden in the clouds, I could have landed early or diverted elsewhere before exposing myself to those unseen hazards. If things had been only slightly worse than forecast and I had to miss the approach at Flying Cloud, I needed to know whether conditions were better nearby, or if I had to proceed east or south to get on the ground. If I knew beforehand, I could avoid a lot of stress and work load when the controller asked, "What are your intentions?" With your knowledge of weather theory, put to use in evaluating aviation weather hazards and prioritizing your goals for the flight, you can be an active participant in the weather process, and not merely be along for the ride.

TRIP TWO: WICHITA TO ADDISON, TEXAS

Occasionally I'm called on to provide pilot service as part of my employment. Usually these trips come with little notice and are fairly urgent in nature. On this occasion I received an 8 a.m. phone call asking if I could transport two executives in a Beech Baron from Wichita Mid-Continent Airport to Addison, Texas, which is on the eastern edge of the Dallas/Fort Worth megalopolis. Figure 15-5 illustrates my route of flight.

It would be a challenging go/no-go decision. It was a rainy, stormy morning in Wichita, and the commercial weather broadcasts had all warned of possible severe weather forming in late morning. I hadn't checked the en route or destination weather yet, but I knew that the Dallas area had suffered strong tornadoes on both of two days prior to the flight. I also knew that this system was the result of a nearly stationary cold front that affected the entire route, intensified by a strong jet stream flowing overhead.

Taking a quick look outside, I answered that I wasn't sure we could make it at the requested 10 a.m. departure time, but I'd check the weather and call back.

Huddled over the computer, I queried DUAT about conditions for a two-hour IFR flight with a 1600Z departure. Reams of paper flowed from the printer; here are the pertinent excerpts, as well as my commentary:

FA HAZARDS AND FLIGHT PRECAUTIONS: current report not available.
FA SYNOPSIS AND VFR CLOUDS/WEATHER:
CHIC FA 271200 COR 2
Synopsis valid until 280400
Clds/wx valid until 272200 OTLK valid 272200-280400
SEE AIRMET SIERRA FOR IFR CONDS AND MTN OBSCN.
TSTMS IMPLY SVR OR GTR TURBC SVR ICG LLWS AND IFR CONDS

NON MSL HGTS DENOTED BY AGL OR CIG.
SYNOPSIS...10Z SFC LOW OVR UPR MI VCNTY SSM WITH CDFRNT EXTNDG SWD ALG SSM-STL-ADM-LBB-ABQ LN WITH 2ND LOW OVR WRN TX VCNTY LBB. FNT WILL MOV EWD.
SERN KS...CLR. 18Z CIG 30-40 BKN-SCT 70 WITH WDLY SCT RW-/TRW-. TSTMS PSBLY SVR. CB TOPS ABV 450. OTLK...MVFR CIG RW TRW.
SRN AND ERN OK...CIG 20 BKN-SCT 80 BKN 120. ISOLD RW~-/TRW- BCMG WDLY SCT RW/TRW BY 16Z WITH TSTMS PSBLY SVR. CB TOPS ABV 450. OTLK...MVFR CIG RW TRW.
N CNTRL AND NERN TX...CIG 20 BKN-SCT 70. TIL 15Z OCNL VSBYS 3-5F. ARND 16Z OVR WRN PTNS AREA WDLY SCT RW/TRW DVLPG SPRDG EWD OVR NERN TX BY 21Z. TSTMS PSBLY SVR. CB TOPS ABV 450. OTLK...VFR N CNTRL TX...VFR RW TRW NERN TX.

Fig. 15-5. *Flight path.*

It looked on paper, then, much better than I expected. Wichita had been experiencing "training" thunderstorms all night, where the upper-level flow pattern brought storm after storm over the same ground track like boxcars passing a railroad crossing. If I could get out between the cells, it seemed, I should be able to negotiate my way around the "widely scattered" storms the area forecasts predicted. Helping me out would be the Baron's airborne weather radar and relatively clear air between the cells, allowing me to visually avoid the heavy stuff.

The printout continued:

SEVERE THUNDERSTORM WATCH NUMBER 145 FOR A LARGE PART OF CENTRAL AND SOUTHERN OKLAHOMA AND EXTREME NORTH CENTRAL TEXAS. VALID UNTIL 200 PM CDT.
THE SEVERE THUNDERSTORM WATCH AREA IS ALONG AND 70 STATUTE MILES NORTH AND SOUTH OF A LINE FROM ALTUS OKLAHOMA TO 80 MILES EAST OF OKLAHOMA CITY OKLAHOMA. A FEW SVR TSTMS WITH HAIL SFC AND ALF TO 3½ IN. EXTRM TURBC AND SUF WND GUSTS TI 80 KTS. A FEW CBS WITH MAX TOPS TO 520. MEAN WIND VECTOR 24040.
SVR TSTM CLUSTER CONTS TO PUNCH ENEWRD INTO CNTL OK.
ACTVTY IS CURRENTLY JUST N OF SFC BNDY AND WRM MOIST VRY UNSTBL AIR OVERRUN BNDRY. XPCD STORMS WILL CONT SVR NEXT SVRL HRS.

Well, they're out there, but still widely scattered. For visual avoidance of the cells, I need to fly high enough to stay above the low-level clouds. It looks like my best choice is to try to stay to the east of the line, and to divert east and land if necessary if things get too close. More DUAT data:

CONVECTIVE SIGMET 36C VALID UNTIL 1655Z. KS OK TX. FROM 60W ICT-30SW OKC-30SE SPS...LINE SVR TSTMS 30 MI WIDE MOVG FRM 2440. TOPS ABV 450. HAIL TO 3½ IN...WIND GUSTS TO 80 KT PSBL...OTLK VALID 271655-272055 FRM SLN-BUM-TXK-DRT-70SE MRF-LBB-GAG-SLN...SVR TSTMS ARE OCRG WRN OK MOVG EWD.

Things are still west of my route, if only I can avoid the smaller cells near Wichita.

AIRMET SIERRA FOR IFR AND MTN OBSCN VALID UNTIL 272000 KS...OCNL CIG/VSBLYS BLO 10/3 IN ST PCPN AND FOG. CONDS DVLPG/SPRDG EWD DURG PD...CONTG BYD 20Z.
AIRMET TANGO FOR TURBC VALID UNTIL 272000 KS...OCNL MDT TURBC 240-430 DUE TO WINDSHEAR INVOF UP LVL JSTRM...ELSW NO SGFNT TURBC XPCD OUTSD CNVTV ACTVTY...TX...OCNL MDT ISLD SVR TURBC BLO 100 DUE TO STG AND GUSTY LOW LVL FLOW ACRS AREA. CONDS CONTG BYD 20Z.
AIRMET ZULU FOR ICG VALID UNTIL 272000 KS...OCNL MDT RIME ICGICIP BTWN 40-80 AND 200. CONDS DVLPG/SPRDG EWD DURG PD...FRZLVL...80-120...OK TX...NO SGNF ICG XPCD OUTSIDE CNVTV ACTVTY.

All right, now I can begin some strategic planning. Assuming again that I can avoid the thunderstorms, if I stay below the freezing level and/or out of the clouds while in Kansas, icing shouldn't be a problem. En route, I don't have to worry much about turbulence, although the call for turbulence at high altitude verified the potential for continued storm development. I'll need to be terribly concerned with that hazard only near the ground near my destination, where shifty surface winds might create a wind shear. Low visibility, the last of the four hazards, shouldn't be a problem near the ground, either. I'll have to shoot an approach into Addison, most likely, but I should break out well above minimums. I watched more information spew from the machine:

SURFACE OBSERVATIONS
ICT (Wichita Mid-Continent Airport) SA 1356 A02A M41 OVC 10+ 184/39/28/0422/007/PK WND 0426.
PNC (Ponca City, in northern Oklahoma) SA 1356 E40 OVC 12 183/44/31/0308G20/008.
OKC (Oklahoma City) SA 1356 A02A M27 OVC+ 156/46/34/0219/001/RB02E33.
ADM (Ardmore, in southern Oklahoma) SA 1347 21 SCT E31 OVC 5F 72/72/1008/989.
ADS (Addison, Texas) 1347 M7 OVC 3F 75/71/1415G25/984/ CIG RGD.
DFW (Dallas/Fort Worth) SP 1419 M16V OVC 6F 1513/984/CIG 14V18.
DAL (Dallas Love Field) SP 1350 MII OVC 4F 76/72/1614G20 /985/CIG RDG.

Most of these reports were about an hour old, and things looked a little darker to the southwest than the Wichita AWOS suggested, but conditions should be marginal VFR to VFR for the entire trip. It was getting a little foggy in the Dallas area, and the ceiling was ragged. The Dallas area commercial airports were providing "special" observations, so conditions might be changing rapidly. I still had "east" as an out if I needed it. There's more:

PILOT REPORTS
ICT UA/OV ICT/TM 1338/FL145/TP BE30/SK 052 OVC 080 145 OVC 197.
OKC UA/OV OKC18030/TM 1328/FL090/TO PA31/SK IMC/TB LT OCNL MDT.
DFW UA/OV DFW/TM 1312/FLDURGC/TP B757/SK 016 OVC 053 CA/ RM A FEW CI ABV.
DAL UA/OV DAL/TM 1354/FLDURGC/TP AR72/SK OVC 055.

Although these also were fairly old, it looked like the area forecast was confirmed, and that flight near Wichita and Dallas above 8000 feet would be in the clear. I'll file for 8000 feet and ask for a climb as necessary in Oklahoma to stay in the clear for thunderstorm avoidance.

RADAR SUMMARIES
ICT 1435 AREA 4TRW++ 216/25 202/45 15W MTS 460 AT
212/46. AREA 1TRW++4R- 319/122 180/117 170W XT 460 AT
237/83.
OKC 1425 AREA 2TRW++IP 332/145 5/120 142/100 191/125
244/155 MT 480 248/38.

Widely scattered thunderstorms were mainly well away from my route of flight but, again, the reports were fairly old.

TERMINAL FORECASTS
ICT AMD 1 15Z C30 BKN 0314G22. 19Z C20BKN 0614 CHC 5R-F.
OKC 14Z C30 BKN 0513 CHC 3TRW. 19Z C35 BKN 0713 OCNL 3TRW.
DFW AMD 1 17Z C25 BKN 1920G30 CHC 2TRW+ G40.
DAL AMD 1 18Z C25 BKN 1920G30 CHC C15 OVC 2TRW+ G40.

Still, the chance of thunderstorms, but comparatively good weather otherwise. Gusty winds in Texas warn again of possible low-level wind shear.

I've already tentatively picked an altitude of 8000 feet to stay above the clouds and to conform to my initial direction of flight, but I want to check the winds aloft in case a little higher works out better. (See Table 15-3.)

Table 15-3. Winds aloft forecast

FT	6000	9000	12000
ICT	1608+10	2119+05	2324–01
OKC	1414+12	1916+07	2124+01
DAL	2228+17	2227+11	2232+04

Yes, it appears I'll have lessened headwinds at a lower altitude, and I'll be out of the prime freezing range of temperatures for the entire trip. I'll still file for 8000 feet.

The DUAT printout offered only one NOTAM that might affect me, that the Addison Brons NDB/ILS LOM was out of service. To review, and to satisfy my goal of safety above all else, here are my decisions concerning the expected weather hazards:

Thunderstorms: They're definitely out there, the most significant hazard for today's trip. I've chosen an altitude that should keep me out of the general scattered to broken layer, high enough to allow visual contact with and avoidance from the cells. My weather radar will be my long-range "eyes" and will provide coverage for the climb and descent through clouds. If conditions deteriorate, I'll turn southeast, away from the line of storms and their general southwest-to-northeast flow.

Turbulence: I expect a little on climb through the clouds, and a lot near the ground in Texas. Most of the way I'll be away from forecast and reported altitudes for turbulence. I'll employ standard wind shear techniques on landing at Addison, and divert to the southeast if winds become excessive.

Reduced visibility: Except on climb and the higher portions of my final descent, visibility shouldn't be a major problem if I avoid the area of thunderstorms. I'll keep my ears open for reduced visibility because of fog in Texas, and I'll divert away from the front, to the southeast, if Addison "socks in."

Ice: The Baron is fuel-injected, so there's no worry of carburetor ice. I've chosen an altitude that will keep me out of the prime icing range.

I could have called Wichita Flight Service for some additional information, but I decided that anything extra they could give would be redundant. I had every reason to believe the air was unstable because big cells were forming, so knowing the precise lifted index didn't matter (I later learned it was –9 to –11). Pilot reports placed the cloud tops for me, so I didn't really need to ask about the constant pressure charts. I filed the flight plan shown in Fig. 15-6, then I called my passengers and asked them not to be late if they wanted to avoid the storms.

Form Approved: OMB No. 2120-0026

U.S. DEPARTMENT OF TRANSPORTATION
FEDERAL AVIATION ADMINISTRATION
FLIGHT PLAN

(FAA USE ONLY) ☐ PILOT BEARING ☐ VNR ☐ STOPOVER | TIME STARTED | SPECIALIST INITIALS

1. TYPE	2. AIRCRAFT IDENTIFICATION	3. AIRCRAFT TYPE/ SPECIAL EQUIPMENT	4. TRUE AIRSPEED	5. DEPARTURE POINT	6. DEPARTURE TIME PROPOSED (Z)	ACTUAL (Z)	7. CRUISING ALTITUDE
VFR							
x IFR	N12345	BE58/A	192 KTS	ICT	1500		8000
DVFR							

8. ROUTE OF FLIGHT
ICT direct ANY V74 Pioneer V77 OKC V163 ADM V358 DFW direct ADS

9. DESTINATION (Name of airport and city)	10. EST. TIME ENROUTE HOURS	MINUTES	11. REMARKS
ADS	2	00	

12. FUEL ON BOARD HOURS	MINUTES	13. ALTERNATE AIRPORT(S)	14. PILOT'S NAME, ADDRESS & TELEPHONE NUMBER & AIRCRAFT HOME BASE	15. NUMBER ABOARD
5	00	X	Thomas Turner on file ICT	3
			17. DESTINATION CONTACT/TELEPHONE (OPTIONAL)	

16. COLOR OF AIRCRAFT
white/brown

CIVIL AIRCRAFT PILOTS. FAR Part 91 requires you file an IFR flight plan to operate under instrument flight rules in controlled airspace. Failure to file could result in a civil penalty not to exceed $1,000 for each violation (Section 901 of the Federal Aviation Act of 1958, as amended). Filing of a VFR flight plan is recommended as a good operating practice. See also Part 99 for requirements concerning DVFR flight plans.

FAA Form 7233-1 (8-82) CLOSE VFR FLIGHT PLAN WITH ________ FSS ON ARRIVAL

Fig. 15-6. *Filed flight plan.*

Ready to go

Around a quarter to ten I called Flight Service from the airport and asked for an abbreviated briefing. I received a repeat of the watches, AIRMETs and SIGMETs I'd seen earlier, and the following:

CURRENT CONDITIONS
"ICT 1520 SP M41 OVC 10+ 38/34/0422/0007/RM TSTM SW."

No real change over what had been forecast.

"OKC 1455 M28 OVC 10+ 46/35/0220/001."

Again, no change.

"ADM 1456 E21 OVC 4F 75/71/1415G22/985/CIG RGD."

Things were happening as forecast!

"ADS 1437 MB OVC 3F 75/71/1415G25/985."

As predicted.

A check of radar showed some widely scattered level four and five cells just southwest of Wichita, drifting to the northeast, and the line, still strengthening, in western Oklahoma and the Texas panhandle. Unfortunately, no new pilot reports had been filed, but I felt I could take off to the west, maybe even northwest, and then decide whether it looked all right to proceed on course once I was above the overcast layer. I amended my flight plan to take off direct to the Hutchinson VOR, which is about 40 miles northwest of Wichita, and then direct to Anthony, and then as filed. I figured this would keep me well clear of the incoming cells while I was penetrating the clouds, and that I could ask for "direct Anthony" as soon as I was in VMC above. If the cells started to close off my route, I'd land at Hutchinson before we got too far away from home. With this added to my other preflight decision-making, we were "go" for an attempt at the trip.

ATC granted my amended clearance. I entered the cloud layer at around 5000 feet and broke through the top shortly thereafter. It was one of those dark-sky days, with a grey cirrus layer above, and ominous, bumpy buildups occasionally poking from the lower layer to that above. Lightning flashed now and then in the big clouds, but the ride was surprisingly smooth.

My route south appeared to be open, so I asked for "direct Anthony" before level-off and was given a vector towards the station. Soon I was level at 8000 feet, in clear, smooth air, indicating 177 knots on the Loran ground speed, about 15 knots' headwind influence.

The trip went along uneventfully, as purple buildups slid slowly by my right wing, and in about 45 minutes I was near the Pioneer VOR, at Ponca City, Oklahoma. I called Flight Watch, filed a pilot report, and asked for updated weather ahead. Nothing I'd been warned of before had been canceled, and there was some new information. An AIRMET for occasional ceilings below 1000 and/or visibilities below 3 miles had been issued for Texas. I confirmed with the briefer that this was due to the fog I'd heard about. Enid, Oklahoma, just west of my route, reported a thunderstorm in progress. That was the buildup I saw to my right. Oklahoma City, on my route 86 miles ahead, also reported a light thundershower in progress, with surface winds gusting to 30 knots from the north. I could see what I assumed was the storm cloud ahead, and I decided to deviate well to the east of the cloud if it was in my way when I got there. Ardmore was still marginal VFR, with a 1700-foot ceiling, while Addison and the other Dallas area airports reported ragged ceilings from 1200 to 1600, and a hot, sticky wind from the southeast. There were several airline-generated pilot reports in the Dallas area, all confirming the ceilings and calling for tops of the first layer below 5000 feet. The ride was uniformly smooth to merely light turbulence during climb through the lower layer. In particular, a Baron had just reported light turbulence below 4000 feet climbing out of my destination airport.

The only potentially ominous note was that both Dallas Love and Dallas Forth Worth were starting to put out amendments to their terminal forecasts. On examination, however, Amendment 2 to each of these forecasts were unchanged except that the gusty winds would be more southerly than earlier predicted. In short, the flight was progressing well in what a physician might call a "guarded" condition. I continued with my eyes open.

When I dialed in the Oklahoma City VOR and identified it, I learned of a Convective SIGMET 37C that had just been issued. Calling Flight Watch, I learned that it called for an area of severe storms from 20 miles east of Salina, Kansas, to Ardmore, Oklahoma, and to the west. The area of storms was moving from the southwest at 40 knots, with maximum tops above 45,000 feet. Hail three-and-a-half inches across, and 80-knot winds, had been reported at the surface.

In other words, that dark sky to my right was worse than it looked. I was still 40 miles from OKC, headed deeper into the advisory area, so I invoked the plan I had chosen before takeoff: divert to the southeast. I called Center and asked for a revised routing "direct present position to Ardmore, then as filed." That would keep me skirting the extreme edge of the convective area, with an eastbound escape route still wide open.

The fine folks at ATC are wonderful about deviations because of the weather. In fact, I've been told one of their worries is that general aviation pilots often don't realize that diversions from an airway are an option. I was quickly granted my request and provided a vector towards Ardmore, Oklahoma, and the stormy skies began a slow retreat behind my right wing.

An hour and a half into my flight, still in smooth air between layers with cumulus clouds poking up to the west, I flew over Ardmore. With around 30 minutes left to go, I checked in with Flight Watch, provided a pilot report, and asked for weather updates. A Center Weather Advisory had just been issued for a line of level 5 thunderstorms, from 30 miles southeast of my position to 40 northwest of DFW, moving northeast at 25 knots. I filed that information for a moment and asked for anything else new. There were pilot reports over Dallas that turbulence was moderate from 8000 to 11,000 feet, and that the lower layer's bases and tops were 2000 and 4500 feet, respectively. The terminal forecasts, still warning of possible thunderstorms and gusty surface winds, were unchanged.

Okay . . . I'm pointed directly at a line of heavy thunderstorms, and I have to fly along the length of their line to make it to my destination. They are moving fairly rapidly to the east and, since I'm so close already, they might catch up with me if I try to parallel them to the east. I asked the briefer if she showed any activity behind the front. Her reply was negative. At least for a hundred miles, there was no thunderstorm activity to the west of the line. That was easy. I reevaluated my preflight decision to deviate to the east and decided it was invalid in this case. I thanked Flight Watch, returned to Center frequency, and received authority to deviate as necessary to the west. For good measure, I descended to 6000 feet, to avoid the reported turbulence ahead.

I was coming up on the Dallas area, and the radio was alive with pilots requesting deviations around individual cells. ATC was doing its best to approve reroutes as necessary, usually granting approval based on see-and-avoid in VMC above the lower layer and between storm clouds. I listened to ATIS at Dallas/Fort Worth, which reported a measured ceiling of 2000 broken, 4500 broken 25,000 overcast, visibility seven, and a gusty southeast wind. Love Field, for comparison, was similar, reporting occasional breaks in the overcast as well. On the limit of reception, Addison reported 1900 broken, 2700 overcast, visibility five miles in fog. The wind was from 150 at 12, gusting to 25 knots, and the altimeter setting was 29.85. I began to configure for the ILS 15 approach into Addison when I was handed off to approach control.

Except for the volume of traffic on the radio, from here in the trip was uneventful. I deviated as necessary around small cumulus clouds that might or might not have been menacing. I watched the radar to make certain there was nothing on the final approach or missed approach courses into Addison. I was kept above the clouds until turning inbound on the localizer for the ILS 15 approach, and I slowed to maneuvering speed just outside of the marker and before entering the murk. It was a bumpy ride for the last three or four minutes, and it wasn't the smoothest landing I've ever made, but we taxied in to the general aviation ramp a little over two hours after leaving a stormy Wichita.

Watching the cable television weather later that afternoon, I realized that our midmorning departure was about the only window we'd have had for making it to Texas that day. Severe storm after severe storm drove across central Oklahoma and into Kansas, blocking any passage south. I confidently made the trip, however, by examining the available weather data, formulating a plan to meet the primary goal of safety, and then obtaining updated information to reevaluate my plan once under way. (Actually, I had two plans, a departure plan to get above the clouds, and then an en route plan.) At each point of my flight I had an "out," an escape route designed to minimize risk, decided on before I ever left the ground. I deviated from that preflight plan only when faced with overwhelming evidence as to why it would not work. Nothing ever came as a surprise. Now let's see how weather conditions can sometimes force a change in plans.

TRIP THREE: CHARLESTON, WEST VIRGINIA to EVANSVILLE, INDIANA

New Year's Day dawned clear and extremely cold as I prepared to fly homeward from Charleston, West Virginia. I planned the first leg to be a leisurely 287 nautical miles westward to Evansville, Indiana, paralleling Interstate 64 along airway Victor 4 (Fig. 15-7). On the trip eastbound, the folks at Evansville took great care of my wife, little boy, and me when we caught up with an ice storm right around sunset, and I knew the flight operations manager at Evansville's Tri-State Aero from when he attended my class in Wichita. It was a good place for a rest stop.

Fig. 15-7. *Planned flight path.*

We were traveling in a rented Cessna 172, very well equipped as the type goes. It had full King IFR, new paint and interior, and a recently overhauled engine that produced a consistent (and admirable) 127-knots true airspeed at 2575 rpm cruise, leaned to best power (100 degrees Fahrenheit rich of peak EGT) at the cost of just under ten gallons per hour.

Arriving at Charleston's Yeager Airport around seven thirty in the morning, we found that the airplane hadn't been hangared overnight, and so the laborious job of deicing began. As the airplane warmed in the FBO's hangar, I called Flight Service for a standard briefing. Here's the gist of what Flight Service had to say:

> FLIGHT PRECAUTIONS: Flight precautions exist for light to moderate mixed icing in clouds and precipitation from the freezing level to 15,000 feet. The freezing level as at or near the surface for the entire route of flight. A SIGMET for light to moderate icing has been posted for areas just north of my route, forecast to

drift slowly southward during the day. Flight precautions also exist for marginal VFR to IFR conditions in the same area as the icing SIGMET.
SYNOPSIS: A quasi-stationary front extends from a low over Lake Erie along a Cleveland-Dayton-St. Louis line, joining with a warm front west of St. Louis towards Kansas City. Motion of the stationary front is slowly southward, to about a Pittsburg-Lexington-St. Louis line by mid-afternoon. Conditions south of the line are generally VFR, and are expected to become marginal VFR to IFR later in the day.
CURRENT CONDITIONS: Charleston, WV: VFR, sky clear and six miles' visibility in fog. Temperature 17, dew point 13, wind light and variable, altimeter 30.03.
Lexington, KY: VFR, ceiling 10,000 broken to overcast, visibility 10. Temperature 25, dew point 20, wind 230 at 10, altimeter 30.00.
Louisville, KY: VFR, ceiling 8000 overcast, visibility 10. Temperature 29, dew point 25, altimeter 29.99.
Evansville, IN: VFR, ceiling 7500 overcast, visibility 7 in haze. Temperature 28, dew point 25, altimeter 29.95.
FORECAST CONDITIONS: Charleston: until 1800Z: 4000 scattered, occasionally ceiling 4000 broken, visibility 7 in light fog and haze. Winds southwest at 10 knots. After 18Z becoming ceiling 2500 broken visibility 5 in fog, occasionally ceiling 1000 overcast visibility 3 in fog and snow showers. Wind becoming northwest at 10 to 15 knots. Outlook: MVFR.
Louisville: until 18Z: ceiling 7000 overcast, occasionally ceiling 5000 overcast, visibility 5 in fog. After 18Z, ceiling 3000 overcast, visibility 5 in fog, occasionally ceiling 1000 overcast visibility 3 in fog and snow showers. Outlook: MVFR.
Evansville: until 18Z: ceiling 5000 overcast, visibility 5, fog; after 18Z, ceiling 4000 overcast, visibility 5 in fog, occasionally ceiling 800 overcast, visibility 3 in fog and snow showers.
Outlook: IFR.

Table 15-4. Winds aloft

	3000	**6000**	**9000**
CHS-LOU	2515	2420	2325
LOU-EVV	2320	2325	2235

NOTAMS: nothing significant affecting my route of flight.
PILOT REPORTS: it was still early, and the only reports were from airliners in the Louisville area. They uniformly called for bases of the overcast in the 5000 foot range (conditions called for only "occasionally" in the forecast), with moderate mixed ice accumulation in the clouds to 9000 feet.

I had already anticipated most of this by watching The Weather Channel the night before. In fact, that's why I bundled up the wife and baby fairly early that morning.

West Virginia had been battered with snow squall after snow squall for the entire duration of our visit, and I knew that we had to get out early that morning if we were likely to make it home anytime within the next few days. We had a small window of opportunity until around noon New Year's Day; if we could make it as far as Evansville, we'd be in clear air all the way to Kansas.

Because I knew the picture was "iffy," however, I was ready with a few more questions for the briefer after he finished his report. As icing was the greatest hazard to a safe journey, the amplifications I requested all concerned ice. The first thing I asked for was for the observed freezing level along my route, which was quickly confirmed to be at or near the surface. Next I asked for the forecast temperatures at 3000 and 6000 feet, and found them also to be just below freezing. I queried for the 850-millibar constant pressure plot. It showed a wide temperature/dew-point spread at roughly 5000 feet most of the way, hinting at clear skies at that altitude. Only the Evansville area called for a temperature/dew-point spread within five degrees, indicating the possibility of clouds near 5000 feet. Finally, I asked for the stability chart lifted index. The index was around –1 south of the front, meaning that the air was a little unstable. That meant cumulus clouds and the possibility of clear icing inside them.

I needed an "out." I asked for weather reports a little further south, in case a reroute was necessary. The forecast didn't change much south to the Tennessee line, except that conditions weren't expected to deteriorate quite as early. A high-pressure system over South Carolina, unfortunately, was pumping moisture over the Blue Ridge, obscuring the passes and creating foggy skies around Nashville and Knoxville. The air wasn't much warmer down there, either. The strong high was what had dropped Charleston's temperature so low overnight, even on the "warm" side of the stationary front. In other words, "south" worked as an escape route only to a point, and I'd be better off to land as soon as practical if conditions got too bad.

Looking back at the "big picture," here's how I evaluated the hazards:

Thunderstorms: no hazard at all. Turbulence: not mentioned in the forecasts, but I'd expect a little turbulence close to the ground the further west I got because of increasingly strong southwesterly winds. I'd be past the mountains before I encountered much of this, so I should be all right as long as I stayed high enough.

Reduced visibility: Except for areas of snow showers, the route should be VFR all the way, with a little fog as snowpack melted with the day's warming. This, of course, assumes I stay south of the front and take off on schedule.

Ice: the "hazard of the day." I have scant little ice protection, so my best bet is to stay clear of all icing potential. I'll go VFR today, to ensure I can maneuver to stay clear of ice-laden clouds, but I'll participate in Flight Following to stay "on frequency" and get real-time information about conditions en route. Many would think an IFR clearance would enhance safety, but in this case it might require me to penetrate clouds or precipitation, which I need to avoid. Clouds are currently no closer than 5000 feet above the ground, so I'll plan on 4500 feet as a cruising altitude. This gets me above the instrument MEA around the West Virginia and Kentucky hills, above the likely low-level turbulence, yet below the base of overcast skies to avoid the ice. I'll employ carburetor heat continuously in cruise to avoid the carb ice hazard, but it's actually almost too cold for carb icing to form.

My "out?" Things should be quite good at least as far as Kentucky, and then I'll have Lexington, Frankfort, Louisville, and all the airports in between if the clouds press too far south. A "180" back towards Huntington, West Virginia, might be a good idea if I need it, and I can head south towards Bowling Green, Kentucky, if all else fails. With all this in mind, I filed the flight plan depicted in Fig. 15-8.

Form Approved: OMB No. 2120-0026

U.S. DEPARTMENT OF TRANSPORTATION FEDERAL AVIATION ADMINISTRATION **FLIGHT PLAN**	(FAA USE ONLY) ☐ PILOT BEARING ☐ VNR ☐ STOPOVER	TIME STARTED	SPECIALIST INITIALS

1. TYPE	2. AIRCRAFT IDENTIFICATION	3. AIRCRAFT TYPE/ SPECIAL EQUIPMENT	4. TRUE AIRSPEED	5. DEPARTURE POINT	6. DEPARTURE TIME PROPOSED (Z)	ACTUAL (Z)	7. CRUISING ALTITUDE
X VFR / IFR / DVFR	N1234Y	C172/A	125 KTS	CHA	1300		4500

8. ROUTE OF FLIGHT: CHS V4 EVV

9. DESTINATION (Name of airport and city)	10. EST. TIME ENROUTE HOURS	MINUTES	11. REMARKS
EVV	2	45	one occupant infant in car seat

12. FUEL ON BOARD HOURS	MINUTES	13. ALTERNATE AIRPORT(S)	14. PILOT'S NAME, ADDRESS & TELEPHONE NUMBER & AIRCRAFT HOME BASE	15. NUMBER ABOARD
5	00	X	Thomas Turner on file ICT FSS	3
			17. DESTINATION CONTACT/TELEPHONE (OPTIONAL)	

16. COLOR OF AIRCRAFT	
red/white/blue	CIVIL AIRCRAFT PILOTS. FAR Part 91 requires you file an IFR flight plan to operate under instrument flight rules in controlled airspace. Failure to file could result in a civil penalty not to exceed $1,000 for each violation (Section 901 of the Federal Aviation Act of 1958, as amended). Filing of a VFR flight plan is recommended as a good operating practice. See also Part 99 for requirements concerning DVFR flight plans.

FAA Form 7233-1 (8-82) CLOSE VFR FLIGHT PLAN WITH ____________ FSS ON ARRIVAL

Fig. 15-8. *Filed flight plan.*

It took a lot longer than expected to deice and completely dry our Cessna, and it was close to 11 a.m. (1600Z) before we finally loaded up to leave Charleston. I tried to get an update from Flight Service as to the conditions en route, but I was placed "on hold" for nearly 20 minutes before I decided my window of opportunity was running out before we ever took off. It was still clear and just a little foggy at the mountaintop airport, so I knew I could at least start westbound and pick up new weather information in the air. I listened to ATIS:

> "Charleston Yeager Airport information Bravo, 1545 zulu observation. Sky clear, visibility 5, fog. Temperature 25, dew point 22, wind 230 at ten, altimeter 30.09. Visual approach runway 23 in use. Advise on initial contact you have information Bravo."

The temperature was a bit warmer than expected, and the wind was a little stronger. This meant more wind from the southwest, itself suggesting that:

- Winds aloft will be stronger than forecast, increasing my time en route.
- Conditions ahead of the front are likely to be a little better than forecast because the southwest was the driest direction right now, and less moisture would therefore be ahead of the front.
- The front was probably moving south faster than expected, with the increase in pressure differential indicated by the breeze. Skeptically, I decided to take off, remembering that a 180-degree turn (with the aid of increased tailwinds) might be my best option if things got worse.

Takeoff was smooth. A ragged, foggy undercast filled the Kanawha River valley beneath. I lined up with Victor 4, asked Charleston Approach to hand me off to Center for flight following, and activated my flight plan.

The first quarter of the journey was beautiful. Bright sunlight reflected off snow-covered hills, endless acres of leafless trees streaking the landscape with red. Far to my right the sky gradually faded to grey, the high overcast lowering at the horizon, where I could see the approaching front. Ground speed hovered around 90 knots, the best I would see all day, but I was thankful that this Skyhawk could outrun even some bigger types.

South of Huntington's Tri-State Airport, I checked in with Flight Watch and filed a "good weather" pilot report: sky clear, flight visibility greater than five miles, negative turbulence, and outside air temperature of +5C. I was particularly interested in the weather ahead. Here's what they had to say:

Lexington was currently 4000 overcast, visibility 5 in fog and light snow flurries. The temperature was 28, with a dew point of 25; winds were southwest at 10 knots, gusting to 15.

Louisville called for ceiling 4000 broken, 7000 overcast, with five miles' visibility in fog and drizzle. The temperature had risen to 33, with a dew point of 28, and the winds were 230 degrees at 15 knots. A Cessna Conquest had reported moderate mixed icing on descent through 9000 feet.

Evansville was currently ceiling 7000 broken, visibility 6 in drizzle, with a temperature of 35 and a dew point of 28. Winds were 240 at 15 gusting to 25 knots.

Just for measure, I asked about Cincinnati's weather, behind the front. With an 800-foot overcast and three miles' visibility in snow and rain showers, it wasn't good. The bad weather was confirmed. South of my route, Bowling Green was still in good shape, with a 10,000-foot, thin overcast and more than seven miles' visibility. South looked good if I needed it, and it was increasingly apparent that I might. I thanked the Flight Watch briefer and started planning a contingency route towards Bowling Green.

In retrospect I might have decided right then to turn southward, but I had a "mind-set" of going on to Evansville, and conditions seemed to be staying VFR as forecast, so I pressed on westbound. Around Lexington, the sky had indeed lowered to a thick layer near 4000 feet. My cruising altitude was just legal at about 500 feet below the bases. Already the radio was lively with reports of ice in the clouds and calls for reroutes, but conditions actually looked better just to the west of the Lexington area. I called approach:

"Lexington, Cessna 1234Y, request." The controller replied. "Lexington," I returned, "Do you mind if I ask anyone on frequency for the bases west of the city?" Pretty soon I had prompted a pilot report from a Metroliner inbound from the west, which indicated a 5000 foot ceiling just west of Lexington, with lower conditions to the east, and "good" inflight visibility. I thanked the Metroliner and the controller and advised that I was going to skirt around the south edge of town, clear of the snow showers over the airport, and reintercept Victor 4 west of Lexington.

Next I called Flight Watch and asked for the latest Louisville weather. Soon I had a report of an 8000-foot ceiling, visibility 5 in fog, and a 20 knot wind out of the southeast. Conditions were better than expected. The western edge of the stationary front might actually be blowing northward as a warm front now, propelled by the strong southerly wind.

Still in smooth air, with better weather ahead, I pressed on, mentally filing "east" as unavailable as an option should conditions deteriorate. "South" was now my escape route. Flight visibility remained good ahead, although the clouds seemed to be lowering once more. Approaching Louisville, however, the bases suddenly forced me down to around 3000 feet to remain well clear of the clouds. Things weren't going as anticipated.

This started to ring some bells in my head. First, I was no longer adhering to the "hemisphere rule" for VFR traffic westbound, and that made a collision more likely. Second, I was on an IFR altitude, however unlikely that an IFR airplane would be on this low altitude for long, and that also warned of potential conflict. There might be other VFR airplanes out here, I thought, and the available airspace is dwindling. I added my landing light to the complement of strobes, nav lights, and beacon. I also reminded myself to be a little more aggressive in looking for airplanes, and asked my wife to watch out from the back seat as well.

I also remembered the huge number of VFR-into-IMC accidents that occur when instrument-rated pilots penetrate the clouds unexpectedly, so I double-checked my IFR en route charts to verify minimum safe altitudes. I was just about ready to ask for vectors for a landing at Louisville when it started to rain.

I looked at my outside air temperature gauge, puzzled that it still read the +5C it had throughout my flight. Suddenly I remembered that I had guarded against the cold by employing a standard Cessna cold-weather trick—taping over the wing root air vents. The Cessna's outside air temperature gauge sits inside the right wing root vent. I had no idea what taping over that hole might do to the calibration of the probe. That was it. It was time to land.

I still had many miles' visibility when I asked Louisville for vectors to a landing at Standiford Field, which was just northwest of my position. Rain drops streaked back from my windscreen, and there was no hint of ice on the wings or struts, but too many variables had been added to the equation, and I felt my best bet would be to land and sort it all out. As if to emphasize the unforecast change in the weather, Approach Control advised me that I had only about 50 knots ground speed on a long final approach to Standiford's runway 29. I touched down on a runway covered with rough patches of ice, although a quick inspection after shutdown showed no ice on the Cessna's airframe.

Should I have done things differently? Perhaps. I could have angled south towards Bowling Green, into the teeth of the southwest gale, and flown around the precipitation that ringed Louisville. Had I done that, I might have continued west a little earlier. As it was, the band of showers quickly cleared Louisville, and I was airborne between a widely scattered layer and a thin, high overcast less than two hours later. What I did do was to get out of Charleston in the only window allowed a nonice-certified airplane for the next several days, and I kept abreast of changing conditions to safely complete at least part of my flight. Once on the ground, I found that conditions were rapidly improving to the west. In fact, had I continued westbound only a dozen or so miles, I would have reached that scattered layer and the clear skies above. Winds had slowed me down to a crawl, however, and I logged exactly three hours from start-up to shutdown, covering only 183 nautical miles along Victor 4. I took the weather information provided by Flight Service, asked pertinent questions to supplement the briefing, and then actively obtained updates en route from a variety of sources to refute the official forecast. I admittedly was running low on options when the rain began over Louisville, but precisely because I no longer had a clear "out," I decided to cut my flight short.

Hopefully, these three examples will suggest how you can critically examine the information you receive from official aviation weather outlets, combine them with nonaviation, commercial weather broadcasts, and make an informed decision aimed at meeting the goals of your flight. Use your knowledge of weather theory and development, as well as those aviation charts and reports that provide additional clues to weather hazards, to ask informed questions that make the go/no-go decision easier. To verify or refute the official forecasts, anticipate the weather phenomena you're likely to see and get updated information if the forecast doesn't come true. Evaluate the weather information you receive based on the four aviation weather hazards (thunderstorms, turbulence, reduced visibility, and ice) and pick a route and altitude designed to limit your exposure to those hazards. Finally, always keep in mind what you'll do if conditions become critical, and land as soon as possible if your options "dry up." If you employ these techniques, you'll be surprised how easily you can contend with most weather situations, and how confident you'll be when making your go/no-go decision.

PART 5

Future weather reporting

16
Postscript

METEOROLOGY IS AN INEXACT SCIENCE, MEANING THE FEWER WEATHER observations forecasters have, and the farther into the future they look, the less likely it is that they'll be able to predict the outcome. Realizing this crucial shortcoming in the way weather forecasts are made, (and realizing the economic and safety impact of less-than-accurate forecasts in aviation as well as other fields) the National Weather Service and other agencies have instituted a plan to increase the frequency and coverage of weather observations.

The goal is to turn more and more away from forecasting as it exists today and enter the realm of "nowcasting," accurately reporting a "snapshot" of weather conditions in thousands of areas on a frequent, regular basis. Weather forecasters will have tremendously greater numbers of observations to provide to users on a real-time basis, and they'll be able to construct much more reliable, if not shorter-term, forecasts. In order to accomplish this economically and with standardization from one reporting point to another, the Weather Service has committed to a network of automated weather observations systems, AWOS and ASOS.

AWOS AND ASOS

Both of these systems, developed in the late 1970s and 1980s, automatically observe weather conditions and report those findings to forecast outlets as well as users through telephone and radio links. AWOS, the Automated Weather Observation System, is the first-generation unit. It was designed to provide reliable, basic weather data at locations not currently served by a human weather observer, thereby increasing the total number of observing points. This was especially helpful to commercial airplane operations; FAR Parts 135 (charter and some commuter) and 121 (airline) require a current weather observation before a pilot can even attempt an instrument approach, so having AWOS on remote airfields made them much more accessible by air.

AWOS observes just the basics: temperature, altimeter settings, and wind conditions. Some versions of AWOS can detect visibilities, ceilings and dew points. The information is synthesized into voice and transmitted over published radio frequencies and telephone numbers (check your Airports/Facilities Directory). In some cases, information is sent via computer keyboards in airport terminals and linked into the forecasting net.

ASOS is the more capable system, able to detect everything an AWOS can observe, as well as amounts and types of precipitation, cloud coverage, and historical trends (peak winds, total rainfall over a period, etc.). ASOS is designed to actually replace human observers at reporting points; it's not uncommon for surface aviation reports and ATIS broadcasts to be the result of an ASOS unit. ASOS observations are taken every six minutes, ten times the frequency of human-based reports, and ASOS data is available in all the same manners as the AWOS information.

As wonderful as these systems seem, they do have drawbacks. Notably, they can record only objective data directly exposed to their sensors. Visibility is only reported in the direction the AWOS or ASOS measures it. A fog bank rolling in to obscure one end of an airport might go unreported. They can only detect clouds passing directly over the sensor, so significant cloudiness might escape their view. Thunderstorms near an airport aren't noticed; hail, tornadoes, and freezing precipitation will not be reported.

Because these automated systems have limitations, and because observations near airports will in the future be based more and more on AWOS and ASOS reports, pilot reports will become even more important to augment the automated report. Get into the habit of noting conditions on takeoff and landing, then filing a PIREP with Flight Service after level-off or landing. That will make these automated measurements more usable and safer for flight planning purposes. You'll come to depend on other pilots to do the same for you.

OTHER UP-AND-COMING TECHNOLOGIES

The other great gap in weather observation and forecasting is the inability to make consistent, objective observations of weather aloft. Currently, sporadic and subjective pilot reports are the only means of constructing a three-dimensional picture of the atmosphere, either for flight planning or forecasting on the whole. New devices are being put into use that promise to allow a ground-based look at conditions aloft on a regular, objective basis.

LIDAR

Light Detection and Ranging is a method proposed to replace radiosondes, or "weather balloons," launched only twice a day and left to blow with the prevailing winds. LIDAR uses a laser beam pointed vertically to accurately track the location and movement of dust and water vapor or crystals at various layers in the atmosphere. Initial tests conducted in the early 1990s are promising; if deemed operational, LIDAR might give pilots their first means of detecting the location and severity of icing and turbulence aloft (based on water droplet size, known temperatures at altitude, and the degree of movement of airborne particles). Pilots will not have to wait for other aviators to encounter hazards and make a report.

Wind Profiler

The Wind Profiler is a system designed to use radar instead of lasers to detect turbulence and moisture aloft. The Wind Profiler is used in testing (and in competition with LIDAR) with good initial results. If either LIDAR or the Wind Profiler is adopted, pilots will enjoy near-real-time reports of turbulence and icing at selected altitudes up to the tropopause.

Radiometers

Radiometers employ a different, unique approach towards detecting moisture and icing aloft: instead of using an active signal, like radar or a laser, the radiometer passively listens for the natural radio frequencies emitted by liquid, solid, and vaporous water aloft. This "Star Trek"-sounding sensor system is still in the early stages of development as of this writing.

Lightning detection

Already in place is a nationwide system of electrical discharge detection systems, sort of a continental Stormscope or Strikefinder that detects the existence, strength, and polarity of lightning strikes. Employed by the Severe Storms Forecast Center in Kansas City, it's used to track thunderstorms from their earliest stage of development through dissipation. When I had the opportunity to watch this at work, the Center workers were just beginning to identify correlations between the polarity of lightning strikes and the stage of storm development, making tracking and short-term predictions much more accurate. This system is vastly superior to normal radars in storm tracking because it can sense the storm through its entire life cycle.

LLWSAS

Another technology being put into widespread use, the Low-Level Wind Shear Alerting System, uses a series of wind speed and direction indicators, placed around an airport, and a central computer that tracks the shifting winds. If one or more sensors differ dramatically from the prevailing winds, or a pattern emerges that warns of a microburst passing near or over the airport, the tower cab is automatically alerted, allowing controllers to pass that warning on to pilots. LLWSAS units are already in place at many of the airline hub airports prone to thunderstorms or microbursts; expect most Class B and C airspace airports to have these warning systems in the future.

Terminal doppler weather radar

Doppler radar can detect not only the presence, but also the relative direction of movement, of areas of precipitation. This provides the same sort of wind shear and microburst information obtained by the LLWSAS, but over a wider area of coverage that extends beyond the airport boundaries. A few units are in service at larger airports in the central United States, where the hazards are more frequently encountered.

NEXRAD

NEXRAD, the Next Generation Radar, is a quantum step forward from even the Terminal Doppler Radar system. It boasts of eight times the resolution and twice the range of existing radars; it can detect and track 16 levels of precipitation as well as determine their movement as a group and individually within areas of storms. This provides a real-time, accurate look at wind movement, pinpointing areas of hail, wind shear, microbursts, and tornadoes. Pointed vertically, future NEXRAD systems might be used to detect winds and turbulence aloft, as well as the precise location of jet streams. The goal of operational NEXRAD units is to provide up to 10 automated observations each hour.

NEXRAD is being installed first in Kansas because of the frequency of the state's severe storms and tornadoes. I've already enjoyed the advantages of NEXRAD technology personally. Last year, soon after the Wichita NEXRAD was declared operational, a tornado warning was issued for the Wichita area. Using NEXRAD, the Weather Service was able to track the precise path the tornado had taken, as well as to predict to the minute when and where it would continue to strike based on the movement of air currents around the funnel cloud. The NEXRAD image was broadcast live on television with a menu of towns and intersections displayed on the screen, including the time at which the tornado would pass each point. Instead of cowering in my basement at the sound of the warning, then, I was able to watch (via my TV set) the storm pass several miles to the south of my home.

Weather uplinks

Mode S transponders have the capability of two-way communications, heralding the day when real-time weather information can be displayed directly in the cockpit. I know of a company that provides this service, using DUAT data, on request using conventional personal pagers; I've seen ads already for a panel-mounted CRT that lists surface observations. I envision a not-too-distant time when you might fly with a panel-mounted moving map overlaid with the location of VFR, MVFR, IFR, and LIFR weather, as well as areas and intensities of thunderstorms, turbulence, reduced visibility aloft, and ice. A profile view on the screen might give those systems' locations and altitudes; as "nowcasting" outlets detect hazards over reporting points at specific altitudes, a central processor will broadcast the information to airplanes en route. A blue-shaded area, for instance, might appear on overhead and profile views to indicate the location of airframe icing; you simply climb, descend or maneuver your airplane around the display to avoid encountering the hazard. VFR pilots could "see" reduced visibility ahead, perhaps shaded in green, and simply fly around the threat.

Truly usable, real-time aviation weather is in its infancy. Until the time when devices like I've described are in widespread use, you'll need to understand what causes aviation weather hazards to form, how they'll move and modify, and what to do if conditions become hazardous. That will enable you to achieve your goals of safety, comfort, convenience, and economy of flight. Weather historically causes 25 percent of all accidents and 40 percent of all fatal accidents in general aviation; knowing how to obtain weather information, including what specialized questions to ask, and comparing that expected weather model to conditions you actually encounter en route, will allow you to verify or refute the required weather briefing and make better decisions that affect the safe outcome of your trip.

Appendix

Self-briefing aviation weather products

AS THE NEED FOR IMPROVED AVIATION WEATHER INFORMATION SERVICES becomes more apparent, more and more government and private ventures have entered the pilot-briefing market. Significantly, the Federal Aviation Administration recognizes most commercial weather vendors as "official" preflight weather briefings, primarily because private services generally take their information directly from government sources.

What follows is a list of the available aviation weather services as of this writing. In making this list I risk omitting other sources of weather information; any such omissions are purely unintentional, and I apologize for leaving anyone out. I created this list by thumbing through the last several months' issues of numerous popular aviation periodicals, pulling company names and phone numbers from those companies' advertising. Any firm I missed simply isn't advertising prominently in the aviation press. Mention of a particular product or service does not constitute an endorsement, and all claims are those of the provider as expressed in advertising.

Pilot self-briefing weather products fall into three broad categories: government-sponsored preflight planning outlets, other preflight planning sources, and airborne or en route weather update systems.

GOVERNMENT-SPONSORED PROGRAMS

The federal government realizes the benefit of providing pilots the ability to self-brief via direct computer access to the Flight Service network. Use of a Direct User Access Terminal (DUAT, for short) is instrumental in the government's plan for consolidating Flight Service Stations and reducing personnel costs. A pilot that self-briefs won't tie up a Flight Service briefer, at least not in making an initial go/no-go decision. That briefer is now more able to provide updates over the radio to aircraft in flight.

Use of the DUAT is free in its most basic form. When querying the computer, a pilot provides the same information required if a telephone briefing takes place: airplane type, departure and destination airports, route and altitude of flight, and estimated time aloft. The computer provides, unedited, all the information the Flight Service specialist would have at his or her disposal. It's up to the pilot to sift through the reams of data to determine any hazards to the flight's safe outcome. To prevent nonaviation users from tying up the lines, access to DUAT is limited to licensed and student pilots only, and the computer automatically cuts offline after 20 minutes, whether the pilot is finished or not. Of course, data can be printed for closer examination or stored on a disk for a more leisurely review.

DUAT in its free form consists entirely of text, with no graphics. The system is capable of decoding the weather hieroglyphics, but that takes more time, limiting the information transmitted in a 20-minute period. There are three DUAT providers, mandated by the government to provide access software free for the asking. For a subscription and user fee, all three also offer "value-added services" such as pictorial graphics and flight-planning services. Contact these companies for information about DUAT access:

Contel DUAT	1-800-767-9989
DTC DUAT	1-800-243-3828
GTE DUAT	1-800-345-3828

OTHER PREFLIGHT PLANNING OUTLETS

If you want services more tailored to your individual needs (for instance, flight planning services with weather input or real-time radar and satellite imagery), there are several outlets that provide those services for a fee. Probably the "high end" of computer weather briefing services is WSI's Weather For Windows, described by *Private Pilot Magazine* as the "Cadillac" of computer weather briefings. Weather For Windows provides just about any weather product or service you might need or want to plan a trip. You can contact WSI at 1-508-670-5000.

Several of the more popular PC-based flight planning programs interface with DUAT data to derive routes, altitudes, and times en route for your particular airplane. Tell the computer where you are, where you're going, and when you want to go there, and it will access your airplane's stored performance data, combine it with an airways database and the DUAT information, and pump out a recommended flight plan based on least total operating cost or least time en route. For more information, contact:

TAU (The Aviator's Utilities)	1-800-767-4828
Flitesoft (RMS Technologies)	1-800-533-3211
Flitestar (Mentor Plus)	1-800-628-4640

Maybe your off ramp to the information superhighway is still under construction, and you don't have a modem-equipped computer handy. Maybe you want to arrange to have weather maps sent to you at a particular time in the future, or at a regular time on a certain day of the week. Perhaps you're out of town on business, staying in a hotel. Weather-by-fax might be for you.

All it takes is a phone call or a faxed-in order, and, of course, the appropriate financial arrangements, and pilot-selected weather graphics will be sent via fax at the appointed time. Most of the information is obtained directly from the Flight Service network, but real-time radar and satellite pictures are available as well. Two services provide standard-style pilot briefings:

JeppFax (Jeppesen, Inc.)	1-800-621-5377
ZFX (ZFX, Inc.)	1-800-876-1232

A third provider, AFSS (American Flight Service Systems, Inc.), provides a "route briefing," weather information for a relatively narrow corridor tailored to your stated route of flight. The intent of this approach is to weed out extraneous information and allow you to focus on your intended route. Call AFSS at 1-602-921-9019.

AIRBORNE WEATHER BRIEFING SYSTEMS

Providing the weather-briefing technology of the future today is ARNAV Systems' MFD 5000. Mounted in the airplane's instrument panel, this "multifunction display" normally monitors airplane performance and operating data. It has the intriguing capability, however, of providing the pilot with near-real-time weather updates on a moving-map display, via an automatic radio uplink developed in conjunction with Pan American Weather Systems. In the weather-display mode, the ARNAV MFD 5000 provides surface aviation observations (the "hourly" reports) and AWOS and ASOS observations, as well as pilot reports, AIRMETs, and SIGMETs that affect the geographic area displayed. It can also overlay radar and satellite graphics to show areas of cloud cover and precipitation. This comes very near the futuristic cockpit weather display I envisioned in this book's last chapter.

To complete the loop, the MFD 5000 also automatically records aircraft identification and type, location (in longitude and latitude), altitude, outside air temperature, humidity, and winds aloft, and radios back this pilot report on a regular basis for inxclusion in the Flight Service briefing network. If you can afford to enter the 21st century, and your panel has the space for the MFD 5000, call ARNAV Systems at 1-206-847-3550.

If you can decipher weather products yourself, and have the proper hardware to receive it, telephone briefings with Flight Service might become a thing of the past for you. Remember, if you have any questions or doubts, or you want more information than your choice of service provides, don't hesitate to call Flight Service for an abbreviated or expanded briefing. Remember also that use of the preflight planning option should be supplemented with information over the radio once en route.

Index

Illustrations are indicated by **boldface** numbers.

L

M

N

O

P

R

S

About the author

Thomas P. Turner has for four years been the lead instructor for Flight Safety International's Beech Bonanza program, and he developed the Aviation Weather Awareness course for that company. Now an account executive for the AOPA Insurance Agency, he is also president of Mastery Flight Training, an aviation safety consulting firm. He is certified to instruct instrument and multiengine students, and he holds a master's degree in aviation safety.

Other Related Titles of Interest

The Pilot's Radio Communications Handbook—4th Edition
Pete Illman
The bible of effective air-to-ground communications—updated to include the latest FAA regulations, airspace reclassifications, and air traffic control procedures. "If you own or are responsible for maintaining a panel, [this book] should have a spot on your bookshelf right next to the airplane's service manual" (Private Pilot).
0-07-031756-9 $17.95 Paper
0-07-031757-7 $29.95 Hard

Avoiding Mid-Air Collisions
Shari Stanford Krause, Ph.D.
Concise, easy-to-understand information on how to steer clear of other aircraft during all phases of flight. A virtual training course—and a unique, integrated approach to a serious safety issue.
0-07-035945-8 $16.95 Paper
0-07-035944-X $27.95 Hard

Aviator's Guide to GPS
BIll Clarke
A practical explanation of aviation's newest navigation system—the Global Positioning System: what it is, what it does, and how to use it. For pilots, hikers, mountain climbers, search and rescue teams, and others.
0-07-011272-X $17.95 Paper
0-07-011271-1 $29.95 Hard

Flying in Adverse Conditions
R. Randall Padfield
A clear, thorough manual of advice for pilots on how to avoid adverse conditions when possible and what to do when such conditions are encountered.
0-07-048140-7 $18.95 Paper
0-07-048139-3 $29.95 Hard

AIM/FAR 1995
TAB/AERO Staff
Packed with more special features (such as page headings to aid in locating data and shaded changes since the previous edition) than any other version, AIM/FAR 1995 puts into one inexpensive source all the up-to-date flight data, procedures, and regulations general aviation pilots, commercial operators and instructors need to fly in the U.S. National Airspace System.
0-07-063084-4 $12.95 Paper
0-07-063083-6 $24.95 Hard

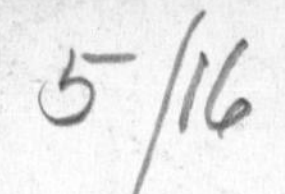